WITHDRAWN

Florida Sinkholes

UNIVERSITY PRESS OF FLORIDA

Florida A&M University, Tallahassee
Florida Atlantic University, Boca Raton
Florida Gulf Coast University, Ft. Myers
Florida International University, Miami
Florida State University, Tallahassee
New College of Florida, Sarasota
University of Central Florida, Orlando
University of Florida, Gainesville
University of North Florida, Jacksonville
University of South Florida, Tampa
University of West Florida, Pensacola

University Press of Florida

Gainesville / Tallahassee / Tampa / Boca Raton / Pensacola / Orlando / Miami / Jacksonville / Ft. Myers / Sarasota

Florida Sinkholes

Science and Policy

Robert Brinkmann

A Florida Quincentennial Book

Copyright 2013 by Robert Brinkmann
All rights reserved
Printed in the United States of America on recycled, acid-free paper

This book may be available in an electronic edition.

18 17 16 15 14 13   6 5 4 3 2 1

Library of Congress Cataloging-in-Publication Data

Brinkmann, Robert, 1961–
Florida sinkholes : science and policy / Robert Brinkmann.
pages cm
Includes bibliographical references and index.
ISBN 978-0-8130-4495-8 (alk. paper)
1. Sinkholes—Florida. 2. Subsidences (Earth movements)—Florida. 3. Geology—Florida. 4. Hydrology, Karst—Florida. I. Title.
GB615.F53B75 2013
551.44'7—dc23   2013020086

The University Press of Florida is the scholarly publishing agency for the State University System of Florida, comprising Florida A&M University, Florida Atlantic University, Florida Gulf Coast University, Florida International University, Florida State University, New College of Florida, University of Central Florida, University of Florida, University of North Florida, University of South Florida, and University of West Florida.

University Press of Florida
15 Northwest 15th Street
Gainesville, FL 32611-2079
http://www.upf.com

To karst researchers and the people of Florida

Contents

List of Illustrations   ix
Acknowledgments   xi

1. Sinkholes in Florida: An Introduction   1
2. Sinkhole Formation   19
3. Geologic Setting and Urban Sinkholes   31
4. Sinkholes in the Ridges and in North and South Florida   72
5. Notable Sinkholes   107
6. Detection and Mapping   134
7. Sinkhole Policy   160
8. Evaluations and Repairs   194
9. Conclusions   217

References   225
Index   237

Illustrations

Figures

1.1. Seffner, Florida, sinkhole   2
1.2. A sinkhole-damaged home   6
1.3. The Winter Park Sinkhole   7
1.4. Florida billboard advertisement for sinkhole help   9
1.5. Florida water management districts   13
1.6. Kentucky sinkhole   15
2.1. Geologic Time Scale   22
2.2. Geologic map of the state of Florida   26
2.3. Diagram of sinkhole forms   29
3.1. Coastal plains of the southeastern United States   33
3.2. Satellite image of Florida   35
3.3. Major ridges of Florida   37
3.4. The Everglades   38
3.5. Map of the Ten Thousand Islands area   39
3.6. Tampa sinkhole swarm   42
3.7. Famous gypsum stack sinkhole collapse   48
3.8. Major sinkholes in Lake County   61
4.1. Spring in the Ocala National Forest   75
4.2. Wakulla Springs in Tallahassee   81
4.3. Lake Jackson and Lake Iamonia   84
4.4. Big Bend region   88
5.1. Paynes Prairie   110
5.2. Devil's Millhopper   116
5.3. Northern Pinellas County   118
5.4. A 1926 sinkhole map of Pinellas County   121
5.5. Sulphur Springs in Tampa   129

5.6. Spring Hill, site of numerous sinkholes   132
6.1. GPR system   139
6.2. Post-development satellite image of northern Pinellas County   148
7.1. Diagram of raveling process   166
7.2. Collapse of a section of the Lee Roy Selmon Expressway   176
7.3. Water-retention pond   184
7.4. Florida cypress dome   187
7.5. Florida cave   191
7.6. Florida cave marked with graffiti   192
8.1. GPR readout showing the location of a sinkhole   205
8.2. GPR operating near a sinkhole   206

Tables

4.1. Tree island classification scheme   105
6.1. Morphometric analysis types   143
7.1. Costs of sinkhole-related losses by year, 1987–1991   172

Acknowledgments

I am grateful to so many people who have helped me create this book. First of all, I owe big thanks to my friend and former colleague Len Vacher of the Department of Geosciences at the University of South Florida, who helped me through early discussions around the organization and content of the book.

Also, I am grateful to many friends, colleagues, and former students who either provided assistance or in some other way helped me through the process. I am sure I unintentionally left someone off of the following list, since this project has been in progress for so very long. If I did not include you, please forgive me. In alphabetical order, I thank John Barco, Todd Chavez, Mick Day, Nicole Elko, Spencer Fleury, Lee Florea, Taiyo Francis, Sandy Gorren, Mario Gomez, Mark Hafen, Matt Harwell, Alexander Klimchouk, Sandra Kling, Sarah Koenig, Melissa Mosquera, Tomohiko Music, Gitfah Niles, Joanne Norris, Leslie North, Bogdan Onac, Mario Parise, Lisa-Marie Pierre, Jason Polk, Phil Reeder, Don Seale, Cindy Shaw, Ann Tihansky, Sam Upchurch, Phil van Beynen, George Veni, Laurie Walker, and Kelly Wilson.

I must recognize here the great contributions of the late Dr. Barry Beck, who did so much to advance the science of Florida sinkholes. His work informed and inspired generations of karst scientists not only in Florida but all over the world.

I am deeply appreciative of Larry Levy, Ina Katz, and Christopher Niedt of the National Center for Suburban Studies at Hofstra University for their support.

Finally, I owe a debt of gratitude to everyone at the University Press of Florida for their kindness and patience.

Sinkholes in Florida

An Introduction

In the winter of 2013 a horrific sinkhole took the life of Jeffrey Bush in Seffner, Florida. He was resting in bed when the ground collapsed under his bedroom. His family rushed in to find just a mattress remaining in the hole where his room once was. Jeffrey was gone forever. One minute he was a vibrant, living human being, and in an instant the earth gave way. This tragedy is unusual. Loss of life to sinkhole formation is a rare occurrence, but sinkhole collapses are not.

In 1993, a large sinkhole opened up on the campus of the University of South Florida in a practice soccer field. Although it didn't make the local news, students, staff, and faculty talked about the presence of the sinkhole on campus and speculated about the overall stability of buildings in the vicinity. The sinkhole, which was 20 feet deep and 15 feet in diameter with vertical walls, formed in the late afternoon, just as daytime students were leaving campus and evening students were arriving. The curious went to look at the sinkhole shortly after it formed, and some brave students jumped into the hole and carved their names in the loose sand on the walls. The next day, the hole was filled and the turf on the field repaired. The sinkhole no longer existed. Shortly after this episode, many students, staff, and faculty began to notice a number of large cracks in some of the older buildings on campus. Few could tell if they were new or if they had formed some time ago. Could the buildings fall into a hole like the one that had formed in the soccer field? Due to the concerns of university people, crack monitors were set on the walls of some of the buildings to determine if the fissures were growing larger. Months later, when it became apparent the cracks were not expanding, the monitors were removed and eventually people on campus forgot about the sinkhole in the soccer field and lost their fear of building collapse (figure 1.1).

Figure 1.1. On February 28, 2013, a sinkhole opened behind the Bush residence in Seffner, Florida, that took the life of Jeffrey Bush. On May 22, 2013, when this image was taken, neighboring homes were being demolished due to extensive damage from the collapse. Hundreds of sinkholes form each year in Florida, but rarely do they cause injury or death.

This sinkhole episode is similar to others that have played out with varying degrees of severity across the state. In Orlando, a sinkhole opened in a mall parking lot to reveal a cavern. During the construction of a large expressway in the Tampa Bay area, a large supporting pier sank 30 feet into a depression. While sleeping at night, a Winter Park woman woke to strange noises and looked out a window to see a sycamore tree disappear into a giant hole. Within a short time, her house followed. In Spring Hill, a foundation cracked from subsurface sinkhole activity, suddenly making a home uninhabitable. These examples represent just a handful of the hundreds and hundreds of sinkholes that occur each year in Florida. Most of the holes are quite small and do not cause significant damage. However, in many areas of the state, sinkholes are all-too-regular events that reduce property values, damage structures, and create headaches for property owners.

Although there are many excellent studies of specific Florida sinkholes, no one to date has put together a book detailing what is broadly known about sinkholes in the state. The purpose of this book is to provide a comprehensive review of sinkhole science and policy in the state of Florida. The topic has received much attention in what is called gray literature, or technical documents, written by consulting firms or by federal, state, and local governments. In addition, several conferences sponsored by the Florida Sinkhole Research Institute in the 1980s and the early 1990s produced excellent volumes of proceedings that remain invaluable sources. There are also many important articles on sinkholes in peer-reviewed journals and books. If I have neglected any particular source of information in this book, I certainly regret the oversight. I hope, however, that the book provides an accurate representation of what is currently known about Florida sinkholes. I believe that the book will be an excellent companion to *The Geology of Florida* (Randazzo and Jones 1997), which mainly focuses on other topics of interest to Florida earth scientists (although Upchurch does excellent service to the topic of sinkholes in his chapter on the environmental geology of the state [Upchurch and Randazzo 1997]).

This book is divided into nine chapters, each of which focuses on a particular aspect of sinkhole science or policy. This introductory chapter summarizes the purpose and contents of the book and discusses how sinkholes are seen in popular culture. It concludes with a review of the current karst and sinkhole expertise in the state and justifies the need for this book. After the introduction, I turn to a detailed examination of sinkhole science. Chapter 2 focuses on sinkhole formation in Florida, setting the stage by reviewing the geology and geomorphology of the state. The chapter also includes a discussion of Florida's hydrogeology, of karst processes and sinkhole formation processes, and a description of the karst landscape of Florida. The chapter concludes with a review of various sinkhole classification schemes utilized in Florida.

Chapters 3 and 4 focus on sinkholes found in two distinct topographic settings in the state. Chapter 3 discusses sinkholes in the urban areas of the Tampa Bay and of Orlando and vicinity, and chapter 4 discusses sinkholes found elsewhere in the state. After the regional overview in chapters 3 and 4, I focus my attention in chapter 5 on notable sinkholes and sinkhole events. Specifically, I discuss the famous Winter Park Sinkhole, Paynes Prairie, Devil's Millhopper, the Pinellas County sinkhole swarms, the Spring Hill sinkhole swarm, and the Sulphur Springs sinkhole swarm. The Winter Park

Sinkhole is notable for its size and the destruction that occurred as a result of its formation. Paynes Prairie, a sinkhole and polje complex located south of Gainesville, is interesting not only from a geologic perspective but also from a historical perspective. Many important events in the history of interaction between Westerners and native Floridians occurred in this area. Located northwest of Gainesville, Devil's Millhopper is an extremely unique sinkhole formation that provides a detailed glimpse into the complex sediments that were deposited above the local limestone. Case studies of the Pinellas County and Spring Hill sinkhole swarms show how sinkholes can impact neighborhoods and local governments. The case studies also demonstrate problems associated with current sinkhole policy. Finally, the Sulphur Springs area of Tampa contains a number of karst features that have been negatively impacted by human activity. Like many areas of Florida, sinkholes in the region are clues to the causes of groundwater and surface-water pollution.

From the discussion of these problematic sinkhole areas, we move to a discussion of sinkhole policy in chapter 6. In recent decades, governments have become more involved in the regulations surrounding sinkholes. One of the best examples of such regulation is the state's definition of a sinkhole for insurance purposes. Floridians may carry sinkhole insurance as part of the standard homeowner's insurance, but there has been considerable debate in the professional community and in the courts as to what constitutes a sinkhole. This chapter reviews the current state of insurance rules and summarizes some of the difficulties faced by homeowners, geotechnical firms, and the insurance industry as a whole as a result of Florida's public policy. The chapter also examines some of the other insurance rules that impact the state's water, caves, and springs.

Because of the intense debate over what a sinkhole is and what it isn't, it is important to understand how sinkholes are mapped. Hence, the focus of chapter 7 is sinkhole mapping and sinkhole investigatory reporting. Research on sinkhole detection and mapping has advanced significantly in recent decades. This chapter reviews state-of-the-art approaches to mapping sinkholes and subsurface voids. This information is important because hundreds of sinkhole insurance claims are made each year in Florida that must be evaluated through mapping techniques. The goal of chapter 7 is therefore to demystify the various approaches to sinkhole detection for the general reader.

Once a sinkhole is found and the funding secured from insurance companies or other sources, any structure involved must be repaired. In chapter

8 I follow the process of handling structural damage, from making an insurance claim to repairing a home. Most homeowners in Florida are not aware of the process they must go through in order to return a home to livable condition. The goal of this chapter is to provide a clear description of what is needed to file an insurance claim and what homeowners should do if a claim is denied. The many approaches people can take to stabilize the ground surface and repair a home are discussed in chapter 8 as well.

The final chapter summarizes the major points of the book and recommends future research directions to better our understanding of sinkholes in the state. While there are scattered efforts underway to study sinkholes and other karst features in Florida, there currently is no organized, statewide effort to coordinate the research. Since the Florida Sinkhole Research Institute was defunded in the early 1990s, there has been no leader in sinkhole or karst research able to bring together karst researchers in the public and private sectors. It is my hope that this situation will be remedied in the future because we need to better understand our karst environment. Some of our biggest challenges in understanding karst and sinkholes in Florida fall within the nexus of science and policy. Thus it is important to foster karst research not only in the sciences but also in tricky policy areas involving important issues such as law, insurance, and karst zoning.

Public Perception of Sinkholes

When the Winter Park Sinkhole formed in 1981, instense media attention and many questions coalesced around the subject of sinkholes (figure 1.3). How do they form? Can they occur anywhere in Florida? What can we do to stop sinkholes? Am I in danger? A feeling of general malaise, somewhat akin to the fear associated with approaching hurricanes, existed throughout the state. Perhaps there was a bit of mass sinkhole anxiety. Individuals expressed concern that their house would be swallowed up by the ground while they were sleeping. Some had nightmares of being trapped inside their home while it was destroyed. The image of the giant Winter Park Sinkhole on the evening news and in newspapers only confirmed public fears of an unstable land that could open up without warning.

This anxiety is related to the fear of nature discussed so well by Schoolman (2001), who explains that the idea of nature as something unknown or frightening is part of the evolutionary development of human thought. Society creates

Figure 1.2. A sinkhole-damaged home in Florida. Photo courtesy of Jason Polk.

an unnatural landscape of luxury homes and shopping malls without the messy intrusion of wild animals or natural disasters, so when nature does invade, it returns us to the instinctive fears deeply rooted in our minds. Just as we have created Bigfoot and the Chupacabra as fanciful expressions of our fear of nature, so too we have developed an unnatural fear of sinkholes. In recent decades, Jeffrey Bush is the only known fatality from sinkhole formation in Florida, even though thousands of sinkholes have formed over this period. Most sinkholes form quite slowly and are subtle features on the landscape. Large sinkholes that form rapidly, like the Winter Park Sinkhole, are rare. But they do occur and thus are real natural hazards that have the potential to cause considerable property damage and loss of life. Yet the reality is that they pose little risk to humans. The greatest risk is to property.

Most sinkhole events have little in common with the Winter Park Sinkhole or the one that took the life of Jeffrey Bush. Instead of a sudden opening in the ground, most of the public are confronted with sinkholes that cause slight damage to homes or property. As we will see in more detail in chapter 2, the reason for this is that most of the sinkholes that form in the state occur as sand filters into buried limestone layers deep underground. As the sand settles into the subsurface, much like sand in an hourglass, a cone of depression forms on the surface landscape and creates a broad, shallow bowl very different in form from the Winter Park Sinkhole.

The shallow sinkholes that commonly form in Florida cause considerable damage to homes each year. Billboards (figure 1.4) and other advertisements that peddle the services of legal firms specializing in sinkhole law or of real estate speculators that purchase sinkhole-damaged homes are ubiquitous in Florida. Such advertisements suggest that when sinkholes damage homes, homeowners get shafted by insurance companies and that it is impossible to be safe in, or to sufficiently and economically repair, a home damaged by sinkhole activity. Nothing could be further from the truth. Most sinkhole claims are settled by insurance companies and, once repaired, sinkhole-damaged homes are not only stable but often retain their value. This is not to say that legal assistance is not needed in some cases or that homes always retain their value after damage occurs. However, the tremendous volume of

Figure 1.3. The Winter Park Sinkhole. Photo courtesy of the *Orlando Sentinel*.

advertising gives the public the perception that sinkholes always do ruinous economic harm and that an attorney should be hired immediately after a sinkhole forms. In this way, the media fosters our exaggerated fear of giant sinkholes forming any time or any place. In turn, our innate fear of nature translates into a fear of economic loss. These two fears—loss of life and loss of property through natural disaster—are powerful indeed. As we will see later on, sinkholes form more regularly in particular areas of the state, and it is quite possible to create risk maps that show where sinkholes are most likely to form. Although these maps can help decrease public anxiety about some areas, they support greater anxiety about others.

In contrast to their fear of sinkhole formation, people have a distinctly different perception of the sinkhole depressions that created wetlands and lakes before the European settlement of the state. While lakes are seen as recreational areas for fishing, boating, and swimming, wetlands and low areas have been traditionally regarded as unwanted landscape features. Prior to the wetlands rules of the 1970s and 1980s, the public was encouraged to fill wetlands to reclaim land for productive uses. Large areas of sinkhole wetlands were lost in the nineteenth and twentieth centuries in Florida as sinkholes and other wet areas were filled for agricultural and urban development. For example, there is evidence that hundreds of sinkholes were lost in Pinellas County, Florida, as it developed in the twentieth century (Wilson 2004). In addition, rural residents in karst areas throughout the United States have used sinkholes as dumps where refuse can be thrown away without blighting the land surface or without going through the effort of creating a local landfill (Dinger and Rebmann 1986) (figure 1.6). Unfortunately, many of these dumps polluted the local aquifers, and today, in some areas of the country, efforts are under way to find these rural dumps and remove the refuse. In Florida, many of the local "grottos"—amateur cavers' clubs—are active in cave conservation efforts including the removal of trash from sinkholes. In cleaning out sinkholes, they hope to protect karst resources and perhaps find new cave entrances. Whatever their goal, the outcome of their efforts enhances the preservation of the karst systems in Florida.

The preservation of wetlands has led to a new perception of wetland sinkholes as parklands. Many sinkholes around the state include hiking trails on their periphery or boardwalks across them. Perhaps the best-known sinkhole and the most hiked sinkhole in the state is Paynes Prairie. This sinkhole group near Gainesville is pivotal to the understanding of the history

Figure 1.4. There are many advertisements for sinkhole attorneys and services on billboards in Florida.

of Florida. Other smaller and lesser known sinkholes can be readily visited within state and national forests and county and local parks. We value these places for several reasons. First, they have tremendous ecological significance as homes to many important plant and animal communities. Indeed, they are refuges for some of the state's threatened and endangered species. Second, they are important water recharge areas. In recent years, many of the state's water management districts and local governments have purchased thousands of acres of sinkholes and their watersheds in order to preserve healthy water supply systems for Florida's various regions. Sinkholes and their natural drainage basins help to maintain aquifer levels and naturally treat water to remove impurities. Third, the natural areas created by sinkholes serve as buffers between natural and urban areas of the state. As population pressures increase in Florida, it is important to maintain some natural areas to prevent Florida's becoming one giant megalopolis of urban sprawl. With coastal areas nearing maximum build-out, development encroaches upon the interior of the state—in the very areas where sinkholes commonly occur. At the same time, many sinkhole-prone natural areas provide contemplative environments away from the stresses of urban living. In recent years, sinkholes in the landscape have come to be regarded as areas to protect and preserve in order to maintain healthy ecosystems.

Sinkhole Expertise in the State

There are a number of sinkhole experts in Florida, but since the early 1990s there has been a distinct lack of coordination of sinkhole research in the state. The greatest volume of research conducted on Florida's sinkholes and karst systems occurred in the 1980s and early 1990s. The start of this period coincided with the formation of the Winter Park Sinkhole, when the University of Central Florida founded the Florida Sinkhole Research Institute, which was partially funded by state resources and partially by grants. Director of the institute, Barry Beck, and his staff worked very hard on a variety of projects, the most important of which was a series of conference proceedings published in the 1980s and 1990s organized around particular themes. The conferences brought together the leading experts in the field to discuss sinkhole research, and the papers gathered in the proceedings make up one of the largest bodies of karst and sinkhole research ever published. The papers are written on a variety of topics, including sinkhole formation, sinkhole policy, and the engineering aspects of sinkholes. The sinkhole conferences have continued into the present era and are now organized by the National Cave and Karst Research Institute.

Another important, long-lasting effort of the sinkhole institute was the development of a sinkhole database. Over the years, the institute kept records of sinkholes that formed in the state. Obviously, they could not track every single sinkhole that formed, because many sinkholes go unreported. (To paraphrase an old saying, if a sinkhole forms in the woods, does anyone hear it drop?) They only sinkholes recorded are those that someone notices, usually because they cause some type of property damage. Obviously, this skews the database to more populated areas. Nevertheless, the database has been used to great success to analyze the formation patterns of sinkholes in the state. Since the institute closed, the Florida Geological Survey (FGS) has maintained this database.

A third important effort of the Florida Sinkhole Research Institute was primary research on sinkholes. Beck and his colleagues conducted a substantial amount of research, which still stands as some of the most significant karst research ever done in Florida. However, the most lasting contributions of the institute are the sinkhole database and the conference proceedings.

Unfortunately, the institute lost its funding in the early 1990s. At that time, Beck moved into the private sector and many of the institute's efforts ceased. Research on sinkholes continues under the direction of Shiou-San

Kuo and focuses mainly on the detection of and on the engineering aspects of sinkholes. However, the Florida Sinkhole Research Institute no longer organizes karst conferences. This is an unfortunate loss to the state, particularly at a time when it is faced with so many sinkhole science and policy issues surrounding the improvement of sinkhole detection and the sorting out of the various insurance complexities confronting homeowners and insurance companies.

Among the karst experts at Florida's universities, the largest group is affiliated with the University of South Florida, where the karst research group includes professors and students from the Departments of Geosciences and Biology. Although relatively new (it formed in 2002), this research group is one of the largest consortiums of karst researchers at any university in the United States. Its expertise base is quite broad, and its members include (1) specialists on the speleology, geology, and geomorphology of karst terrain; (2) experts on using karst features to interpret past climate change; (3) experts on human land use and karst; (4) experts on sinkholes; (5) experts on geophysical techniques used to detect karst features; (6) experts who study the organisms and minerals associated with particular karst features; and (7) researchers who study the hydrology of Florida karst. There are also karst experts working in the University of South Florida College of Marine Science. Florida State University's Oceanography Department conducts varied research on the karst and limestone features found underwater, and the university's Center for Insurance Research has been active in investigating insurance issues associated with sinkhole formation. Geologists at the University of Florida are also active in karst research. In short, scientists and researchers at Florida's universities keep karst research alive and current.

An important source of karst expertise in the state is the Florida Geological Survey. This organization has a number of karst experts on staff who study aspects of Florida karst including karst landforms, karst stratigraphy, and karst hydrogeology. Founded in 1907, the FGS is part of the Department of Environmental Protection. According to the organization's website (http://www.dep.state.fl.us/geology/default.htm/), the FGS is

> the only agency within Florida state government that provides this [geological] information and interpretive data dissemination and necessary institutional memory to support the need for geology and earth science related information to: government agencies, land-use planners,

environmental and engineering consultants, mineral owners and exploration companies, industry, and the public. Program outreach, regarding earth science education and the pre-historic development of our state is also provided to the public and educators.

A quick review of the FGS website indicates that sinkholes are a significant focus of their effort. Indeed, the site is a wonderful resource for the public. It links to frequently asked sinkhole questions and provides excellent answers. The site also links to individuals employed by FGS who have distinct expertise in karst science. In addition, it offers FGS-published reports and maps for download.

The FGS also maintains a sinkhole database and provides online forms for reporting the formation of a sinkhole. This report form requests a great deal of information, including the date, time, and location of the formation, how long it took for the sinkhole to form, the damage caused by it, and the land use and topography in the vicinity of the sinkhole. The maintenance of this database is a wonderful service to the people of Florida because long-term monitoring of sinkholes helps researchers discern patterns that assist in their understanding of sinkhole formation and the geographic variations informing sinkhole risk.

The Department of Environmental Protection, the agency that hosts the FGS, also carries out a variety of research projects on issues related to karst and sinkhole science, including the study and documentation of the flora and fauna found in sinkholes. Other state agencies, such as the Florida Department of Financial Services, conduct research on sinkholes, although such research is often not the main mission of these organizations. The Florida Department of Financial Services, for example, reviews insurance policies associated with sinkhole formation and has a sinkhole insurance expert on staff who serves as a public resource.

The state's water management districts provide significant expertise on sinkholes and water resource issues surrounding them. Florida is divided into five distinct water management districts: the Northwest Florida Water Management District, the Suwannee Water Management District, the St. Johns Water Management District, the Southwest Florida Water Management District, and the South Florida Water Management District (figure 1.5). Each district is charged with the management and protection of surfacewater and groundwater resources. They are also responsible for flood management, land purchases for protection of water resources, protection of resources in times of drought, and regulation of water consumption by overseeing consumptive use and well permits. In order to conduct their work, it is crucial that these organizations have karst experts on staff. Indeed, groundwater hydrologists, due to the nature of the Florida aquifer systems, are experts on Florida karst. Thus many of the staff at the water management districts are among the state's karst experts, and many of the reports generated by the water management districts contain a great deal of information on Florida karst.

Local governments also conduct karst research, albeit from a more applied perspective. Often, karst research is done under the guise of engineering projects, such as the building of a particular structure. Subsurface investigations, which are standard practices in the construction of many large buildings in

Figure 1.5. Extent and boundaries of Florida's water management districts.

Florida, can reveal interesting aspects of the karst terrain. Much local information is contained in the so-called gray literature, not easily accessible to the public, but local government officials, often housed within an engineering or planning department, can be extremely helpful in obtaining local documents of interest to karst researchers. Many staff in other local offices are knowledgeable about karst systems in the region, too.

A number of private companies specialize in sinkholes as well. While I don't want to name any one company at the exclusion of others, suffice it to say that there are dozens of such firms in the state. Most of them work at trying to determine for insurance purposes whether structural damage is the result of sinkhole activity. Others keep karst experts on staff in order to better understand subsurface water flow. These firms create mountains of technical reports each year, many of which are public documents. This information can be useful in understanding sinkholes and karst systems, although locating it can be a daunting task.

Another important source of karst information in the state comes from local grotto organizations. In areas of the United States where caves are found, caving clubs, called grottos, are common. Loosely affiliated with the National Speleological Society, a national organization that focuses part of its efforts on cave conservation, grottos vary in their local missions, but they often conduct cave explorations and help to educate the public about cave conservation. In Florida and other states, many grottos get involved with cleaning sinkholes that in the past were used as dumps. It isn't uncommon, for example, in rural and suburban areas for people to use sinkholes as dumps for old farm equipment or household waste. With time, the sinkhole fills with enough debris to create a potential source of pollution to the subsurface aquifers. Grottos remove this kind of refuse material, thereby reducing the pollution threat. On occasion, they find cave entrances at the bases of sinkholes. Because of their efforts, grotto members often know a great deal about the nature of sinkholes in their region and are wonderful sources of information.

Of course, local residents know their terrain. Individuals who have lived in a particular area for years or whose family has been settled in a locale for generations are often storehouses of knowledge about the landscape. They can point out where sinkholes originally formed, and many remember the damage caused by them. Thus, while we have many experts in Florida who know a great deal about sinkholes, we should recall that local residents can provide unique insights through their direct experiences.

Figure 1.6. Debris at the bottom of a Kentucky sinkhole. Photo courtesy of Jason Polk.

A few years ago, I was teaching a geomorphology class at the University of South Florida that focused on the earth's landforms and how they came to be. About midsemester, the class got to the two-week section on karst landforms. After giving the final lecture on the topic, a student came up to me and brought several pictures. As it turned out, after the first day of the karst section, a sinkhole opened up in front of her condominium between the condominium building and its parking lot. The sinkhole took out part of the parking lot and the student's car sank into the void. Luckily, a tow truck was able to pull her car out of the hole and there was minimal damage. However, the event brought a change in the student's focus. She was a very good student before the event, but after the formation of the sinkhole, she focused like a laser beam on the topic. The same kind of interest often seizes the general public. When individuals are personally affected by a sinkhole, they develop a thirst for knowledge. They seek out information on sinkholes and how they form. With this in mind, I hope this book will help inform the public about the geology of sinkholes and how sinkholes impact us in Florida.

The Need for Karst Knowledge

Nowhere else in the world are so many people touched by sinkholes as in Florida. The state's unique environment is prone to instability and sinkhole formation. Weekly, at least one sinkhole forms somewhere in the state and does some type of property damage. Although we hear the most about the larger sinkholes, hundreds of sinkholes form in Florida each year that we never hear about and that never make the news. Given their frequency, it is surprising how little we know about sinkholes. Researchers have been studying sinkholes for decades in the state, and we have learned a great deal about their formation and their distribution. Indeed, the Florida Sinkhole Research Institute did a terrific job putting together information that was available at the time and brought together sinkhole experts from all over the world. Yet since the early 1990s, there has not been a coordinated effort to study Florida sinkholes or to provide a forum for bringing together the region's experts to talk about their research or set a research agenda.

This recent lack of coordination is unfortunate. There are arguably more sinkhole problems today than there were back in the 1980s and early 1990s. These problems can be divided into two areas: science and policy. Science and policy problems cannot be completely separated because science impacts policy and policy issues create scientific questions. It is useful at this point to examine some of the key scientific and policy research questions that remain elusive to us in Florida.

Science Questions

> *How do sinkholes form?* Sinkholes form in a number of different ways in Florida. Many researchers have examined sinkhole formation and have found that local geologic and hydrologic conditions greatly influence how sinkholes form in any given area. Yet, as we will see, there are patterns as to how sinkholes form in a particular locale.
>
> *Why do sinkholes form?* There seem to be two main causes of sinkhole formation in Florida: natural and human. We know for certain that sinkholes formed in Florida long before humans significantly altered the landscape through modern development and massive water withdrawals. We also know that the processes that acted to create sinkholes in the past are still in action today. Thus, through the rule of uniformitarianism (the present is the key to the past), we know

sinkholes will continue to form through natural causes in the state. As we will see, there seems to be some seasonality to sinkhole formation. Climate variations can also impact formation timing. Even so, we are a long way from being able to predict when sinkholes will form. We also know that humans can induce sinkhole formation by pumping water from the subsurface or by adding loads (such as buildings) to the land surface. Still, we are unable to predict what specific altered conditions lead to sinkhole formation.

Where do sinkholes form? There are trends of sinkhole formation in Florida. We can map broad areas where sinkholes are likely to form because of specific geologic conditions. We can also see trends in the clustering of sinkholes within these areas. However, researchers are unable to predict exactly where sinkholes will form with any degree of certainty or specificity. At this time, only broad statements regarding regional risk can be made.

Policy Questions

What type of insurance rule is most appropriate for protecting homeowners? In the 1980s Florida required insurance writers to cover property losses from sinkhole formation as part of standard homeowners' insurance policies. As we will explore later in detail, the state, in legislatively requiring insurance coverage, provided an ambiguous definition of the term "sinkhole," creating a great deal of confusion for insurance companies, homeowners, and the courts. Because of this, a number of consulting firms were founded to evaluate what is and what isn't sinkhole damage. Since then, the insurance issue seems to be constantly in flux at the state level, and it is likely that additional changes will occur in the future. However, it is still not clear what the best situation for homeowners is. Also, the state does not want to create a situation in which insurers are financially unable to operate in Florida due to losses incurred by sinkhole formation.

What is the impact of sinkholes on real estate? One of the more interesting phenomena that has developed as a result of sinkhole damage to hundreds of homes each year is the growth of the sinkhole real estate speculating industry. One cannot travel major roads in Florida without seeing billboards that announce "We Will Buy Your

Sinkhole-Damaged Home Today." At this time, we do not understand how this industry affects real estate value. Do homeowners recoup their losses by selling to these organizations? How are neighborhoods affected when a home or several homes are damaged by sinkholes? Does the overall value of property in the neighborhoods decline? Also, are homes damaged by sinkholes repaired effectively? These questions are compounded by the current vexing economic downturn and concomitant property devaluations.

Are those responsible for inducing sinkhole formation responsible for the damages? The question of who is responsible for sinkholes that form as a result of human action has not yet been tackled by the courts. In later discussions, we will consider a number of sinkholes that formed as a result of human activity. In all of these cases, however, a void space was already present under the ground surface and the sinkhole would have eventually opened up at some point in the future anyway. Can someone really be held liable for damages incurred when a sinkhole occurs at a date earlier than would have occurred naturally?

•

Clearly there are research questions that demand some degree of assessment. I hope this book provides a base of knowledge that will help advance sinkhole science and policy. As I worked on this book and talked to friends about it, many of them told me that they were very interested in sinkholes because they thought they had a sinkhole forming in their backyard or in their neighborhood. Almost everyone I talk to, in fact, has a sinkhole story (this is not surprising considering the large number of sinkholes that form each year), though most people do not understand sinkhole science. In the next chapter, I will review how sinkholes form in the state and provide a geologic framework for understanding their formation and distribution.

Sinkhole Formation

The formation of sinkholes in Florida is a common phenomenon brought about because of collapsing void space in the subsurface. But what causes the voids? What causes the collapse? And what is the subsurface like in Florida? These questions are the focus of this chapter. None of them can be answered effectively without reviewing some of the fundamental concepts important to understanding Florida geology. Although at first glance, the state's geology is simple, with stacks of horizontal bedrock covered by Quaternary marine sands, the truth is that although much of Florida's bedrock is relatively young, its geologic history is complex and just now beginning to be understood.

After reviewing some basic concepts of geology and karst processes, this chapter will review Florida's geologic history and stratigraphy, discuss specific karst processes, and conclude by summarizing the principal sinkhole types found in the state.

Basic Concepts

The earth's crust is a solid body made up largely of rocks and minerals. Some of the more common minerals in the crust are quartz, olivine, hornblende, biotite, and feldspar. Rocks are aggregates of one or more minerals. In Florida, one of the most common rocks present in the state is limestone, which consists largely of the mineral calcite, although impurities such as quartz and clays are common accessory minerals. Calcite is a mineral that is commonly found in the bones of animals and in the exoskeletons of crustaceans. When the animals or crustaceans die, their bones or exoskeletons help to make up limestone rock. In addition, fine-grained limestone can form from calcite-rich mud made up of the excrement of small marine life. Calcite can also precipitate out of calcium-rich saline waters. Regardless of the specific formation process, limestone forms in aquatic environments and it is commonly fossiliferous.

As limestone forms, it is laid down in horizontal beds following the geologic principle of original horizontality, which defines sedimentary rocks, such as limestone, as forming grain by grain in horizontal deposits. After their formation, the rocks may become tilted through folding or faulting. However, the original deposition is horizontal. This principle is important in helping to understand the formation history of rocks. In Florida much of the limestone is slightly tilted, perhaps in response to the predetermined form of the Florida Arch and subsequent minor tectonic warping (Beck 1986).

Another principle important to Florida geology is the law of superposition, which states that rocks are laid down in temporal sequence. What this means is that older rocks lie beneath younger rocks. Of course, this assumes that the landscape has not been significantly folded or faulted to cause younger rocks to slide beneath older rocks. In Florida, where the structural postdepositional tectonic changes of the bedrock are minor, the law of superposition holds quite nicely. In fact, the idea of a "pancake" stratigraphy—horizontal layers of bedrock that are progressively older with increasing depth—is an apt conceptual model for visualizing the bedrock geology of the state.

Another key idea is stratigraphy, which is the study of rock strata. Geologists have formalized the way rock strata are defined, mapped, and delineated, and the rules that have been agreed on for stratigraphic studies are found within the *Stratigraphic Code* (North American Commission on Stratigraphic Nomenclature 2005). The *Code* provides a comprehensive guideline for researchers on how to map and define rock strata. In Florida, groups of rock strata have been named and mapped as have rocks throughout the world. Of course, it is quite difficult to map rock strata in Florida because there is so little rock exposed in the state due to the lack of topographic expression, the lush subtropical vegetation, and the subtle cover of marine sand on top of the rocks in many portions of the state. Thus much of what we know about the subsurface in Florida comes from the study of geologic cores that were collected in the process of drilling wells or during mining and quarrying prospecting. We also know a great deal about the subsurface due to geophysical studies. Over the years, as knowledge has come forward about the subsurface, what we know about Florida stratigraphy has increased. As stratigraphers study the bedrock, they create a stratigraphic column that shows the name and type of rock strata found in particular areas. As a result, today we have a detailed stratigraphic column for many portions of the state.

The stratigraphic code requires that rocks be divided into lithostratigraphic units called formations. Formations are groups of similar rocks defined by lithology. The formations are given names that correspond to the location where the best example of the formation is exposed or present in the subsurface. In Florida, for example, the Ocala Formation is so named because the best exposure of the rock sequence is found near the city of Ocala, Florida. The named formations correspond to a particular geologic time as well. Thus particular formations are associated with a particular age. The dating of the rocks, placing them within a specific time period, can be done through relative age dating techniques such as fossil identification or through absolute age dating techniques using one of the many geochemical techniques that date rock formations. What matters for the purpose of this book, however, is that the rocks we will be talking about have been studied thoroughly and have been mapped and named by professional geologists in consultation with their peers. The formations that are identified in Florida are precisely defined.

Time Scale of Florida Geology

Geologic time is vast. The earth is approximately five billion years old. During those billions of years, continents formed, crustal plates moved across the planet through plate tectonics, the sea level rose and fell, the atmosphere formed, and a vast array of geologic changes occurred. What we know of the record of the earth has been encompassed within the Geologic Time Scale, a graph that shows the ages of distinct changes and events in planetary history. Most of geologic time is taken up by the Precambrian eon, a time period lasting from the formation of the earth to approximately 550 million years ago (figure 2.1). The best evidence of the Precambrian eon exists within old areas of the continents called cratons, the "cores" of continents on which newer material was accreted. North America's craton is located in central and eastern Canada and the northern fringe of the U.S. Great Lakes states and portions of New England. There is no Precambrian-aged rock near the surface in Florida. During the Precambrian, rudimentary life forms developed, such as single-celled organisms.

The Paleozoic era, which extends from the close of the Precambrian to the start of the Mesozoic (approximately 240 million years ago), is known for the development of complex marine life. There are many deposits of Paleozoic-aged rocks around the world. In North America, for example, there are many

Eon	Era	Period		Epoch	Age in Millions of Years Before the Present
Phanerozoic	Cenozoic	Quaternary		Holocene	0.01
				Pleistocene	1.6
		Tertiary	Neogene	Pliocene	5.3
				Miocene	23.7
			Paleogene	Oligocene	36.6
				Eocene	57.8
				Paleocene	66.4
	Mesozoic	Cretaceous			144
		Jurassic			208
		Triassic			245
	Paleozoic	Permian			266
		Pennsylvanian			320
		Mississippian			360
		Devonian			408
		Silurian			438
		Ordovician			505
		Cambrian			570
Precambrian	Proterozoic				2500
	Archean				3600
	Hadean				4550

Figure 2.1. The Geologic Time Scale.

Paleozoic sedimentary rocks that serve as the bedrock foundation for the upper Mississippi Valley and Great Lakes region. The Niagara Dolomite is one of the most dramatic formations in this area. It is a Silurian carbonate rock that is prone to karst processes. When exposed, it creates a prominent escarpment that serves as the shelf from which Niagara Falls descends. Of course, there are other, less dramatic examples of Paleozoic-aged rock in the world, however, none exists near the surface in Florida (though examples of Paleozoic rock have been collected from boreholes deep beneath the surface of the peninsula).

The Mesozoic era is also poorly expressed in the state. It represents a time period when life was progressing from sea to land. Lasting from the close of the Paleozoic until the start of the Cenozoic era (approximately 60 million years ago), it is an age in which plants diversified across the continents and large reptiles, the dinosaurs, ruled. The Mesozoic was also a time in which large deposits of organic matter were preserved in swampy environments. Thus Mesozoic-aged sedimentary rocks are best known for their coal and petroleum potential, which derived from these deposits. The large coal deposits in the Appalachian Mountains and Rocky Mountains are within Mesozoic formations. Likewise, most of the oil that we use originated in living organisms that saw the light of a Mesozoic sun. Florida has no Mesozoic rocks exposed at the surface, although some have been found in deep cores drilled within the state.

The rocks and sediments found at the surface in Florida all formed during the Cenozoic era. This relatively short era, lasting only for the last 60 million years, is a time that is well represented within Florida's stratigraphic record. During this time, the planet saw the expansion of the diversity of mammals, many of which evolved during the Mesozoic. Florida Cenozoic rocks contain some of the best fossil examples of the types of mammals that lived during this era. For those interested in the topic, a visit to the Florida Natural History Museum in Gainesville is in order. Here you will see very nice displays of intriguing animals that once roamed Florida—the giant sloth and the saber toothed tiger among them.

Another place to see great fossils is the Mulberry Phosphate Museum in the small mining community of Mulberry in Polk County. This museum is dedicated to the history of the phosphate industry in the region and to the understanding of the importance of phosphate mining in the United States. Florida is home to one of the richest phosphate deposits in the world, found in the Bone Valley Formation. The word "bone" in the name is an indicator of the richness of the fossils in this formation. Indeed, much of the phosphate mined in the Bone Valley Formation occurs in the form of the mineral apatite within the fossilized bones of Cenozoic creatures such as mastodons, whales, and dugongs. The museum houses a nice collection of fossils, and it often places piles of phosphate ore outside of the museum from which one can collect the fossilized bones of Cenozoic animals.

The Cenozoic rocks in Florida also contain evidence of changing environmental conditions within the state. The carbonate rocks all formed within shallow warm seas, but there is evidence that there were periods during which the rocks were exposed. Karst processes that create voids that lead to sinkholes usually are active only when the rocks are not saturated by seawater. As the sea level rose and fell throughout the Cenozoic, there were karst landscape and sinkhole formation events when sea level was low and carbonate rock formation events when sea level was high. Along with the carbonate rock formation events, there were transgressive and regressive siliciclastic shore deposits that blanketed much of the state as shorelines moved when the sea level changed. The sands that blanket much of the state were deposited during the most recent sea level high.

In older maps, the Cenozoic era is divided into the Tertiary period and Quaternary period. The Tertiary was the longer of the two periods, representing most of the Cenozoic, except for the last 1.8 million years. This period was broken into four epochs (in order of age): the Paleocene, the Eocene, the Oligocene, and the Miocene. The last 1.8 million years are within the time represented by the Quaternary period. The Quaternary is of much interest to geologists because this is the time of the great ice ages, known as the Pleistocene epoch, and of the current postglacial age known as the Holocene. Because geologists have learned so much about the Cenozoic era in recent decades, the time period has been reevaluated and new divisions were created. Today the Cenozoic is divided into two periods: the Paleogene, lasting from approximately 60 million years ago until approximately 20 million years ago, and the Neogene, lasting from 20 million years ago until the present (figure 2.1). Most of the published maps used to represent Florida's geology use the older terminology, and I have chosen to utilize these terms for the sake of consistency.

Geologic Map

The stratigraphy of Florida of interest to karst scientists is the near-surface limestone rock layers and the sediments that cover these limestone layers. The geologic map published by the state of Florida (http://www.dep.state.fl.us/geology/gisdatamaps/ms146_geology_of_fl.pdf/) shows a variety of rock types present at the surface that formed within the Cenozoic era.

The Panhandle

The major rocks and sediments within the Panhandle are the Oligocene Suwannee Limestone near the Big Bend area and the Miocene Alum Bluff Group and Torreya Formation in much of the eastern Panhandle. These deposits are all loosely consolidated sediments consisting largely of sands and clays. They are chronologically the same as the Hawthorne formation in southern and western Florida. The Pliocene Citronelle Formation in the western Panhandle differs in that it contains a greater abundance of clay. These Miocene and Pliocene sediments overlie older limestone, such as the Suwannee Limestone. Fluvial sediments derived from the rivers draining the piedmont areas to the north are found within river valleys and within the delta region of the Apalachicola River.

Peninsular Florida

The peninsula of the state is characterized by a series of north-to-south trending parallel ridges separated by valleys in the northern half of the region and by broad lowlands in the southern half. The ridges are cored with Ocala (Eocene) or Suwannee (Oligocene) Limestone and may be covered with younger sediments. East of the main ridge area and between ridges are found a complex of Miocene and Pliocene deposits known as the Cypresshead Formation and Hawthorne Group. The Bone Valley Member of the Peace River Formation, a part of the Hawthorne Group, is located east and south of Tampa and is the major phosphate deposit in the state. As in the Panhandle, these younger Miocene and Pliocene deposits overlie older limestone in many areas. Quaternary sediments, deposited during the most recent sea-level rise or by wind in the form of dunes, are found specifically along the coastline in barrier islands, along the western edge of the ridges in the form of ancient dunes, or in the southern peninsula in the form of sands, silts, clays, and recent fossiliferous limestone. This younger, Quaternary limestone can be affected by karst processes, but the principal rocks involved in karst landscape formation in the state are the Ocala and Suwannee Limestone. The Avon Park Formation, which is older than the Ocala Limestone, is rarely exposed. However, it too is subject to karstification, and its collapse can lead to sinkholes. It underlies much of peninsular Florida.

It is important to note that the indurated, or hard, limestone in which voids and caverns form is largely covered in the state. Based on the map in figure 2.2 (for a more detailed map to accompany this text, please visit the Florida Geological Survey's website at http://www.dep.state.fl.us/geology/gisdatamaps/state_geo_map.htm), it is evident that there are very few places where the Suwannee or Ocala Limestone is exposed. Instead, most of the

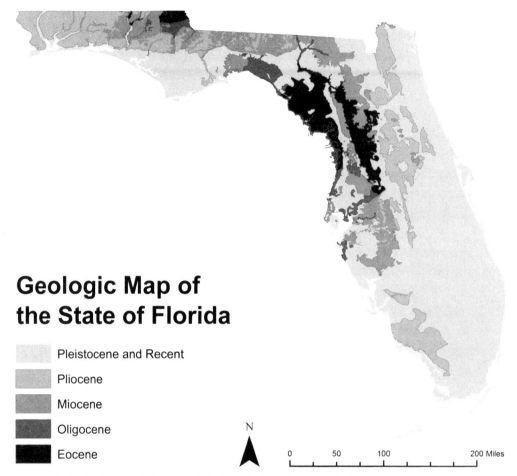

Figure 2.2. Geologic map of the state of Florida. Modified from the Florida Geological Survey.

limestone is covered with some thickness of sediment, and it is the presence or absence of sediment that determines the type of sinkhole that forms in any given area. The cross sections in the online figure demonstrate that the limestone is nearest the surface in the northern portion of the peninsula and in the central portion of the panhandle. As we will see, the thickness of the cover above the limestone greatly influences the distribution of sinkholes in the state.

Sinkhole Types and Definitions

The term *sinkhole* is used in this book to mean a type of depression caused by the solution of limestone. The term is not a particularly scientific one. Karst scientists around the world have given their other, regional names to the form. Terms such as *swallet, doline, dolina,* and *uvala* all have some specific meaning for distinctive sinkholes around the world. However, in Florida, the term that has been used widely for the karst depressions that exist throughout the state is sinkhole. Although the more scientific term is doline, the average citizen in the state is much better acquainted with the term sinkhole, and thus that is the word used in this book. Certainly there are some in the scientific community who would rather Floridians get familiar with the term doline, but this term remains uncommon in the lexicon in the region.

The use of the term sinkhole is somewhat problematic in that it is a generic term used in a variety of settings. For example, the word is used to represent depressions formed from the washout created by a water main break or by a subsidence depression caused by earthquakes. It is not used exclusively within the context of karst depressions. Nevertheless, the vast majority of the public is not familiar with the more exotic names, so the more generic name must do in this context. It is important to note that Florida, and much of the United States, is not alone in its use of terms for dolines that are unique to specific regions. Speleogenesis' online "Glossary of Cave and Karst Terms" notes that a variety of terms, many of them regional, are synonyms for the word sinkhole: jama; pit; ponor; sink; sinkhole; stream sink; swallet; swallow hole; sumidero. Then, there are the non-English synonyms: (French) doline; (German) Dolinen, Karsttrichter; (Greek) tholene; (Italian) dolina, pozzo naturale; (Russian) karstovaja voronka, karstovaja kotlovina; (Spanish) dolina; (Turkish) duden, kokurdan, huni; and (Yugoslavian) vrtaca. Thus the sinkhole of Florida is the tholene of Greece. However, all are considered dolines, a much more respectable and accepted term.

In Florida, the karst is considered eogenetic, which means that it forms in rocks that have not been deeply buried. This is in contrast to the karst in other parts of the world, where the karst forms in limestone that has been buried and lithified (telogenetic karst). There are many differences between eogenetic and telogenetic karst, the most striking of which is perhaps the porosity of the rocks. In eogenetic rocks, there is a high primary porosity because the rocks have not undergone significant compression and lithification due to burial. In contrast, telogenetic rocks have significantly lower porosities. The expression of this difference can be seen, for example, in the hydrographs of springs. In telogenetic karst, the springs react to individual rainfall events because the main connections between large pores are the karst voids and there is little primary porosity remaining in the bedrock that influences groundwater flow. In eogenetic karst, as in Florida, the karst voids are connected with primary pores within the bedrock. Thus spring flow doesn't have the same type of reaction as those springs located in areas of telogenetic karst (Florea and Vacher 2005). Instead, the eogenetic bedrock stores much water within its pores, making reaction time to individual storm events muted or nonexistent.

The difference between sinkholes in telogenetic karst and eogenetic karst is also notable. The density of sinkholes in places such as Florida is very high. There are many small to large sinkholes across the landscape. There seems to be no limit to where sinkholes can form and to their distribution within the Florida karst plain. In contrast, sinkholes in telogenetic karst areas are less widespread. There simply are fewer voids where sinkholes can occur in many telogenetic karst areas.

There are a number of classification systems for sinkholes (see, for example, Waltham et al. 2005; Ford and Williams 1989; and Gutiérrez et al. 2008) that have identified a number of different types of sinkhole forms. In Florida, the karst occurs in areas where bedrock is exposed at the surface and in areas where the bedrock is covered. Sinkholes that form directly on bedrock are called solution sinkholes. In these areas, water collects in depressions. With time, the water can aggressively dissolve the bedrock from the surface downward to create distinct depressions on the landscape. Solution sinkholes are not widely distributed across the state. Most are located on the west coast of Florida on a strip of land where bedrock is near or at the surface. There are a few other areas, particularly on the surfaces of the ridges, where these solution features exist. In addition, some are found within the Everglades. However, it is important to note that most of the karst in Florida is found where the bedrock is covered with some type of sediment.

A covered karst landscape is defined as a karst landscape that develops where the rock is covered by some type of sediment. In Florida, marine sediments consisting of sands and clays mantle the karst. Where the cover is thick, the sediment covers any previous karst landscape that would have formed during low sea level events. In addition, the formation of any depressions after the deposition of thick sediment layers is rare. In these places, such as in Manatee, Palm Beach, and Sarasota Counties, the landscape contains few karst features. In contrast, where the sediment cover is thin, some paleokarst features, though muted, can be identified, particularly within and adjacent to the ridges in the state. However, many karst depressions formed in the state since the deposition of the marine sediment during high water stands.

There are two main sinkhole types that form in a covered karst landscape: cover collapse and cover subsidence (figure 2.3). Cover-collapse sinkholes are the most dramatic sinkholes in that they form rapidly. In contrast, cover-subsidence sinkholes form slowly. These sinkholes are much more common than the cover-collapse sinkholes.

The cover-subsidence sinkholes occur as sediment slowly ravels, or filters, into voids within the bedrock. With time, the slow raveling process will create a surface depression like that seen within an hourglass. Initially these depressions are often difficult to discern due to their subtle nature on the landscape and the low relief of the form. Yet as time progresses, they can grow into very large features. In comparison, the cover-collapse sinkholes are quite dramatic. They form suddenly and create relatively steep-walled depressions. Shortly after their formation, the walls of these sinkholes collapse to create a gently sloping depression that may cause the feature to be indistinguishable from a cover-subsidence sinkhole.

Cover-subsidence and cover-collapse sinkholes can occur almost anywhere in the state where there is limestone beneath sediment, but the majority of them occur within the portion of the peninsula that is north of Lake Okeechobee. There are two main areas where sinkhole activity is most active in the state—the Tampa Bay area and the Orlando-Gainesville-Ocala region. The reason for this activity is the nature of the subsurface. The presence of eogenetic rocks with a high primary porosity makes karstification rapid in these areas compared with other areas of the world. In addition, the warm temperatures and high rainfall amounts make a perfect chemical environment for the dissolution of the bedrock. The control of the distribution is not fully understood, but there is certainly a relationship between modern sinkhole activity and the depth of cover over bedrock (Tihanksy 1999). In areas where the cover is thinnest, sinkhole formation is more common.

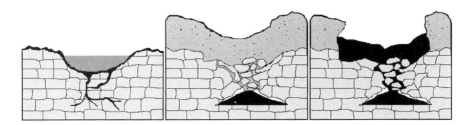

Figure 2.3. Sinkhole forms (*left to right*): (1) solution, (2) cover subsidence, and (3) cover collapse. Modified from Tihansky (1999).

It must be noted that all of these cover-collapse sinkholes form in relation to high permeability paths in subsurface bedrock, often in caves. Thus the study of speleogenesis, or cave formation, becomes an important ancillary aspect of understanding sinkholes. There are a number of detailed studies that describe the complex nature of how caves form (Klimchouk et al. 2000; Klimchouk 2007; Palmer 2007). Both speleogenesis and the morphology of cave systems greatly influence the distribution and characteristics of sinkholes in the state.

The next chapter focuses on distinct sinkhole regions and notable sinkholes that have formed in the state.

3

Geologic Setting and Urban Sinkholes

When examining the hydrology of Florida, one is struck by its vast number of lakes and wetlands, particularly in the peninsula. The distribution, size, and orientation of these bodies of water reveal a great deal about the geologic history of Florida and the formation of the state's karst terrain. Most of these lakes and wetlands are sinkholes and have their origins in the dissolution of limestone and subsequent subsidence. This chapter discusses the distribution of sinkholes in the state of Florida and highlights some particular sinkholes that are interesting because of the way they formed or because the research that has been conducted on them in recent years reveals something interesting about the geology of Florida.

Descriptions of spectacular geologic wonders in other parts of the world, such as the Grand Canyon or Mount Lassen, can include ground-level images that give us a sense of the topographic variations in the landscape. However, when it comes to discussing sinkholes in Florida, aerial photos and maps are usually the best choice for giving readers a visual sense of the landscape. Therefore, in this chapter, I liberally employ maps and aerial photos to give readers a visual context through which to understand the discussion. Many maps inadequately show the true distribution of sinkholes in the state because they lack detailed contour intervals that delineate subtle depression features. Thus aerial photos are perhaps the best way to demonstrate the extent of the distribution of sinkholes in the state.

Another issue at stake in examining sinkholes is determining the size at which to begin mapping them. Is there a particular size at which sinkholes are defined as such? Is there a specific depth at which a karst depression is defined as a sinkhole? Our science has not tackled these questions yet, and the size at which we should begin to map sinkholes is open for debate. Certainly, regional studies only include the largest sinkholes that show up in aerial photos or on topographic maps, but in local studies, particularly in studies of foundation failures or other property damage, much smaller sinkholes can be mapped. Some of these smaller sinkholes can continue to grow or they can end up as ephemeral features on the Florida landscape by filling with sediment through slope wash.

There are a variety of scales for examining sinkholes in Florida. Patterns exist at the state, regional, and local levels. Many of these patterns are visually fractal in the sense that patterns observable on the larger state level can be seen repeating at the regional and local levels as well. Yet what controls these patterns? Clearly subsurface geology is the dominant factor in the distribution of depressions across the landscape, although existing geomorphology and surface geology also impact the distribution of sinkholes.

The size, shape, and depth of depressions also vary considerably across the state. The distribution of variation in these characteristics again reveals that subsurface geology is crucial to the size, shape, and depth of sinkholes. However, surface geology and geomorphology also are important in the formation of these attributes.

Our land uses, too, have had a major impact on sinkholes, particularly in agricultural and urban areas. We have filled sinkholes in and developed housing projects on top of them in the urban areas, and we have filled them in and created agricultural fields on top of others in the rural areas of the state. As a result, many current aerial photos provide only a glimpse of the surviving sinkholes in the state. As places such as St. Petersburg, Tampa, Ocala, and Orlando grew in the twentieth century, housing developments sprang up around the larger sinkholes while more troublesome depressions such as wetlands or seasonally wet lowlands were filled to allow for development. Nevertheless, the existing record of sinkholes as identified by large lakes and depressions is quite helpful in determining the significance of karst processes in the development of the geology and geomorphology of the state.

Geomorphology of Florida

Before discussing the sinkhole regions of the state, it is important to place this work in the context of the broader geomorphology of the region. On a continental scale, Florida is wholly within the coastal plains physiographic province (figure 3.1), a region that in the eastern half of the United States extends in a broad swath from Texas to Florida on the Gulf Coast and from Florida to New England on the Atlantic Coast. This region is largely a depositional

Figure 3.1. The coastal plains of Florida formed from the deposition of marine sediment via coastal processes as sea level rose and fell several times over millions of years.

coastline that has expanded through sedimentation from the breakdown of the once-larger Appalachian Mountains to the north.

There is a general sense that the coastal plain is featureless. In comparison with the more dramatic physiographic provinces of the eastern United States, such as the Appalachian Mountains and the Ozark Plateau, that would seem to be the case. However, there is tremendous diversity within all segments of the coastal plain that allow division of the province into subregions that can help us better understand the geologic history of the East Coast. Because much of this province formed in the late Cenozoic, our current knowledge about this region provides important clues to our recent earth history.

White (1970) divides the coastal lowland plains of Florida into three broad subregions. He maps a Southern or Distal Zone dominated by the Everglades lowlands, a Central or Mid-Peninsular Zone dominated by parallel ridges and intervening lowlands, and a Northern or Proximal Zone dominated by the Apalachicola delta and highlands bordering Georgia and Alabama. Schmidt (1997), in a map that is modified from White et al. (1964) and White (1970), further divides the state into five physiographic provinces: the Coastal Lowland, which rims nearly the entire state; the Central Highlands, encompassing the ridge country of the interior peninsula; the Tallahassee Hills north of the Tallahassee area; the Marianna Lowlands in the north-central portion of the Panhandle; and the Western Highlands north of Pensacola in the Panhandle.

Schmidt (1997) further breaks down these regions in a map modified from Cooke (1939) in which the broad regions are broken into much smaller local areas. For example, Schmidt divides the Southern or Distal Zone into the Everglades, the Atlantic Coastal Ridge, the Southern Slope, the Coastal Swamps, the Southwestern Slope, the Cypress Spur, the Eastern Valley and the Caloosahatchee Valley. The multiple and varied regions listed above certainly suggest the great diversity of Florida's coastal plain. North of the Southern or Distal Zone, the landscape is divided into lowlands, ridges, and uplands. Here, the ridges parallel the north-south trend of the peninsula. The most northern region is divided into lowlands, plains, highlands, ridges, and slopes. In the Panhandle, the landscape no longer trends north-south but follows an east-west trend roughly parallel with the northern Gulf Coast shoreline.

Sinkholes can happen anywhere in Florida, as the entire state is fully underlain by limestone, but the topographic ridges, valleys, highlands, and uplands have a particular impact on the distribution and size of sinkholes.

Sinkhole Distribution on a State Scale

When examining a satellite image of the state of Florida (figure 3.2), one is struck by the patches of blue that pop out from a verdant landscape. Most of these patches are large sinkholes. Lake Okeechobee, the largest of the lakes, stands out as the largest interior water feature in the state and, indeed, as one of the largest inland lakes in the United States. North of Lake Okeechobee, the state is dominated by two distinct lake districts. One of these, the Orlando Lake District, consists of numerous sinkhole lakes extending from Lake Kissimmee in the south to the Ocala National Forest in the north and from Lake Jesup in the east to the Green Swamp in the west. This district is home to one of the fastest growing populations in the country, with most of the development focused around the appealing shorelines of the recreationally active sinkhole lakes. West of the Orlando Lake District is the Tampa Lake District, which extends from Tampa in the south to Masaryktown in the north, and from the Green Swamp in the east to near the Pinellas County line in the west. Like the

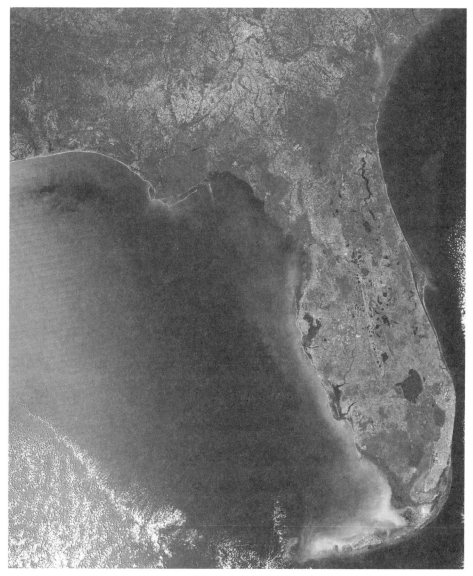

Figure 3.2 Satellite image of the state of Florida. Note the many sinkholes visible in the image.

Orlando Lake District, this area is also experiencing tremendous population pressure.

Prior to the wetland laws that were implemented in the 1970s, many of the sinkholes in the Orlando and Tampa areas were converted into developable land. The conversion often took the form of agricultural land improvement to allow farmers to utilize the maximum acreage of their landholdings in crops such as citrus or tomatoes. It is difficult to drive farm machinery around subtle wetlands, and it was seen as a prudent and economical move to fill in the low depressions. As urbanization advanced into the agricultural fringes surrounding Tampa and Orlando, a great deal of farmland was converted to residential or commercial land use. Thus, many structures in these areas sit on top of former small wetlands or lakes. It is the existing sinkholes, typically in the form of lakes, that provide an overview of the regional characteristics of the sinkhole swarms in these regions.

When examining the satellite image of Florida in figure 3.2, it is easy to see the linear pattern of sinkhole distribution. Paralleling the peninsula, the sinkholes occur along clear lines that extend roughly north to south. This trend parallels many landscape features in the state. Indeed, there are several ridges, such as the Brooksville Ridge and the Lake Wales Ridge, that have the same trend (figure 3.3). Also, paleo-shorelines that rim the peninsula all roughly parallel the north-south orientation. It is not surprising, therefore, to discover that the distribution of karst landforms, such as sinkholes, also have a dominant, approximate north-south peninsular trend. The exact cause of this trend will be discussed in detail when I discuss the Tampa and Orlando sinkhole swarms, however, the distribution is largely caused by joint patterns in the subsurface bedrock, sea level changes and orientations, ridge morphology, and the overall depth of the surface cover of unconsolidated Pleistocene sediment.

Another interesting aspect of the distribution of sinkholes in the state as seen in figure 3.2 is the relative paucity of sinkholes in the Panhandle, the areas south of Lake Okeechobee, the areas north of the Tampa and Orlando sinkhole swarms in the peninsula, and the Big Bend region. The geomorphology of the panhandle of the state is complex. It consists of older piedmont formations in the extreme north, such as the Tallahassee Hills, coastal deposits, and fluvial deposits, particularly the large delta landscape created by the Apalachicola River. Nevertheless, karst features are present. Indeed, the state's only cavern open to public viewing is present in Florida Caverns State Park near Marianna. Here, and in many other areas of the Panhandle, karst features, including sinkholes, are common. Generally, they are not filled with water and thus do not show up on satellite images.

The areas south of Lake Okeechobee have very young limestone near the surface and have not had time to develop extensive sinkholes. Nevertheless, shallow depressions have formed on the surface of the bedrock, and these have allowed cypress domes to form within the marsh (figure 3.4). Thus the sinkholes in the Everglades are not present in satellite or aerial photos as blue

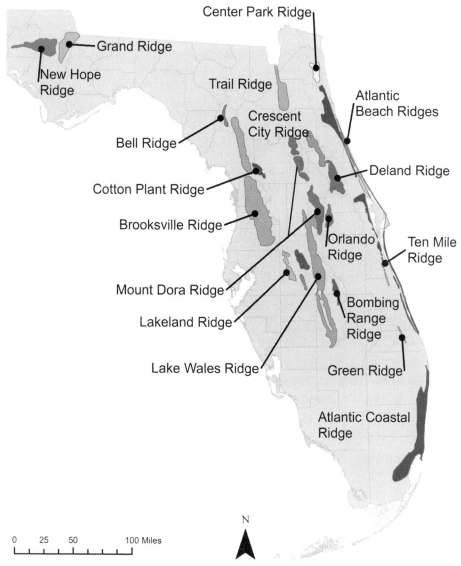

Figure 3.3. The major ridges of Florida.

features but as darker green ovoid shapes surrounded by a lighter green continuous plain. The sinkhole form is elongated due to the drainage flow of the Everglades. Water flows south in a radial pattern from Lake Okeechobee (Schmidt 1997).

The area north of the Orlando and Tampa sinkhole swarms also contains karst features, although like the karst areas of the Panhandle, they are not always found to contain water. Here, we find the karstic Brooksville Ridge

Figure 3.4. The Everglades. Cypress domes are common in some portions of this vast wetland.

and the Ocala Ridge. These two ridges contain some of the oldest bedrock exposed in the peninsula (Ocala Limestone and Suwannee Limestone). Many of the karst features found in these areas are older than the karstic features forming in the Orlando and Tampa areas. Indeed, many of the exposed portions of the ridges clearly indicate—because the rock is no longer saturated and the karst formations developed long ago—that the region consists of a great deal of paleokarst. Interestingly, many of the largest springs in the state are found within or on the fringes of these ridges. Rainbow Springs, Crystal Springs, and Juniper Springs are all good examples. The Big Bend region of the state is also karstic, even if its lakes cannot be seen from space. Here, as in the Everglades, bedrock is very near the surface. In many places, bedrock is exposed to reveal a cairn karst structure, such as the area around the Ten Thousand Islands. In maps, the complex karstic shoreline is clearly evident (figure 3.5).

It must be noted that we are just now beginning to learn about the distribution of karst features underground. Geophysical techniques allow us to ascertain the distribution of void space underground, but we do not have any clear statewide maps showing the characteristics of this world. Lane (1993) notes

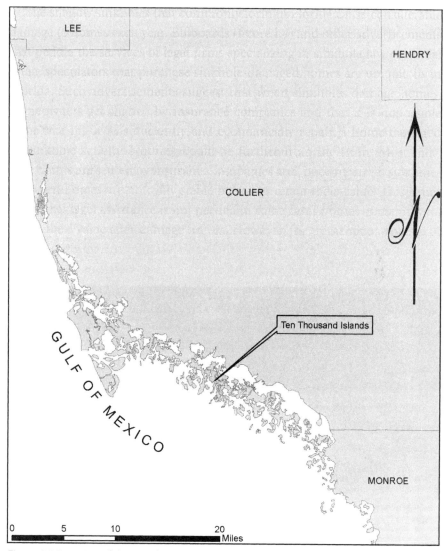

Figure 3.5. Location of the Ten Thousand Islands area of southwest Florida.

that underground caverns in Florida are extensive and rival the extent of void space in notable caves such as the Mammoth Cave system in Kentucky. Much of what we are learning about this underground world comes from intrepid cave divers who risk their lives to expand our knowledge of underground karst systems. What they have found reveals that the state has numerous caverns under the ground surface, many of which are interconnected. Each of the caves has the potential to collapse and cause a concomitant drop in elevation at the land surface to create a sinkhole.

We are also just beginning to learn about the distribution of recently formed sinkholes. The Florida Geological Survey published the Florida Sinkhole Index (Spencer and Lane 1995), which lists approximately 1,900 sinkholes that have formed in the last thirty years. The list is broken down by county, and the sinkholes are located by latitude, longitude, township, and range. The FGS updates the list, and up-to-date information is available on its website. This list is used extensively by scientists to ascertain the distribution of sinkholes in the area. Unfortunately, the list does not include many sinkholes that go unreported, and thus it is just a sampling of the more significant sinkholes that have been reported to officials.

Schmidt and Scott (1984) note that there are two broad sinkhole regions in the state of Florida: a region in north Florida associated with the Chattahoochee Anticline and a region associated with the ridge country of central Florida. They further define these areas as being conterminous with the Floridan Aquifer in locations where it is at or near the surface. At the crests of these topographic features, the authors note, the aquifer can be either unconfined or confined, and the thickness of the sediments on top of the bedrock increases away from the crests. While the observation that there are two broad regions of sinkhole formation in the state is accurate, it must be noted that sinkholes occur in most portions of the state and therefore their distribution can be geographically described by state sinkhole subregions. Thus the discussion of sinkholes of Florida can be divided into the following regions: the Tampa sinkhole swarm, the Orlando sinkhole swarm, sinkholes of the Brooksville Ridge and Ocala Ridge, sinkholes of north Florida, and sinkholes of south Florida.

Before discussing particular sinkhole regions, it is important to note that there are several areas of Florida where sinkholes are rare. As noted earlier, sinkholes can occur anywhere in the state, yet there are some regions in the state where geological conditions make their formation a rare event. Schmidt and Scott (1984) discuss several of these locations. The first of these is in the western Panhandle. Here the limestone in which voids often form is located many tens of feet beneath a thick body of sediment called the Citronelle Formation. This area will be discussed in more detail in the section on north Florida sinkholes, as there are sinkhole-like depressions in that area that were probably not caused by the collapse of limestone. In this part of the western highlands around Pensacola, sinkhole development from the collapse of limestone is unlikely to occur because of the great depth of cover.

Another area where sinkholes are unlikely to form is the Apalachicola delta region. Because this area has been undergoing sedimentary deltaic deposition for thousands of years, hundreds of meters of sediment have been deposited on top of the limestone bedrock. Due to the great thickness of sediment cover, there is very little chance for surface expression of any karst collapse event.

The extreme northeastern portion of the state is another zone where sinkhole formation is unlikely. This area, known as the Jacksonville basin, also has a thick cover of sediment—in this case Miocene carbonates and clastic sediments—over the limestone in which we see the greatest development of karstic void spaces. Schmidt and Scott (1984) note that some small sinkholes do form in this area, but they are caused by the dissolution of thin carbonate shell beds in the surface sedimentary body.

The last area reported by Schmidt and Scott (1984) as having limited potential for karst development is in south Florida in what is called the South Florida Basin, where slow subsidence occurred from the Jurassic to Middle Miocene. The subsidence allowed a tremendous amount of sediment to accumulate in the basin. Among the deposits since the Middle Miocene are a variety of marine deposits, including limestone and sand. Although it is true that sinkholes in this region are rare, it must be noted that they do occur and are important for local ecology and drainage. Most of this area's sinkholes occur in the more recently deposited limestone that is at the surface. Thus, the distribution of surficial limestone is a critical factor in the location of sinkholes in this region.

When these areas (the western Panhandle, the Apalachicola delta, the Jacksonville basin, and the South Florida Basin) are excluded, what remains is an extensive sinkhole zone that extends from the ridge lands west of Lake Okeechobee north to the state line and from just east of Orlando to the Gulf Coast. These areas contain the greatest concentration of sinkholes and also have the greatest potential for sinkhole development in the state of Florida.

The Tampa Sinkhole Swarm

The Tampa sinkhole swarm exists in a region extending roughly north and south from Tampa to the southern edge of the Brooksville Ridge near San Antonio. Parallel to this area is another related swarm of sinkholes in Pinellas County, across Tampa Bay from the Tampa sinkhole swarm. The Pinellas County sinkholes are fundamentally different from the Tampa sinkholes in that they are smaller and are not usually filled with open water.

42   *Florida Sinkholes: Science and Policy*

The Tampa sinkhole swarm is easily identifiable on a map (figure 3.6) as a series of lakes ranging in size from quite small to more than a square kilometer in area. The largest of these lakes, such as Lake Carol in the Carolwood area of Tampa, are surrounded by urban development. Indeed, the northern portion of the sinkhole swarm in Pasco County has seen tremendous growth in the Lutz and Land O' Lakes areas.

As noted above, the sinkholes in the Pinellas County area are different

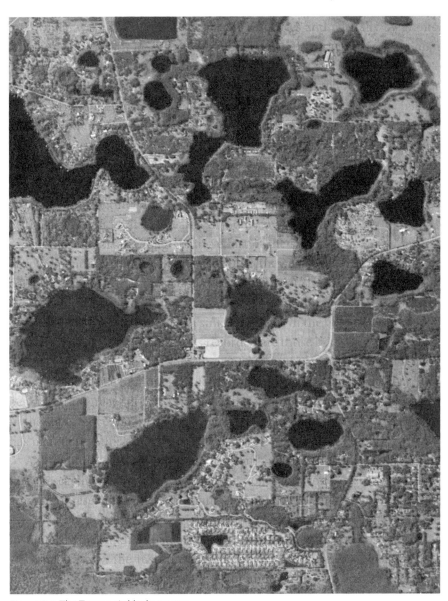

Figure 3.6. The Tampa sinkhole swarm.

from those in the Tampa sinkhole swarm. Unfortunately, many of the Pinellas sinkholes have been destroyed as the county has grown into the most densely populated county in Florida. Using unique 1920 aerial photos, Wilson (2004) studied the distribution of sinkholes in Pinellas County prior to urbanization and found that hundreds of sinkholes were filled as development progressed. She also found that there are very few large sinkhole lakes similar to those found in the Tampa swarm. Also, the northern part of the county contains complex karst depression features that contain wetland vegetation. They may have formed from a combination of karst landscape and dune field development. Many of the sinkholes were lost as urbanization progressed through the county. There is a notable lack of sinkholes present in the area of the City of St. Petersburg in the 1920 photos, probably due to the urbanized nature of that part of the county at that time.

The most comprehensive publication describing the formation processes involved in the Tampa sinkhole swarm and other sinkholes throughout west-central Florida is by Ann Tihansky (1999) of the United States Geological Survey (USGS). Tihansky maps sinkholes that formed from 1960 to 1991. Most of these newly formed sinkholes were found in areas where there was a thick, clay-rich cover. Here, rapidly forming cover-collapse sinkholes occur with some frequency, and numerous sinkholes also form in areas where there is thick permeable sand. Tihansky points out, however, that the types of sinkholes that form in these areas are the cover-subsidence kind while cover-collapse sinkholes are typically induced by human activity. In areas where the cover is thin and in areas where the cover is over 60 meters thick, few sinkholes form. Thus, according to Tihansky, an important factor in the distribution of currently reported sinkholes in west-central Florida is the thickness and type of sediment above bedrock. She also notes that sinkholes in rural areas are underreported. As so much of Florida is undergoing rapid urbanization, it is likely that we will see reports throughout the region increase.

As Tihansky points out, there are a number of cases where human activity spurred sinkhole formation. Major inducers of sinkholes are the large well fields mined to produce drinking water for the region. For example, she describes a particular case study describing the Section 21 Well Field in Hillsborough County, noting that as pumping increased significantly, sinkhole formation increased concomitantly. She found this to be true for other well fields that came online over the last several decades. In addition to drinking-water well fields, wells used to protect crops from freezes can also be problematic. The Tampa Bay region is one of the nation's most productive agricultural areas, particularly for citrus, strawberries, tomatoes, truck produce, and landscape plants. When temperatures fall to near freezing, farmers in the region spray water pumped from the Floridan Aquifer onto their farm plants. The temperature of this water is near 23 degrees Celsius, and thus it protects the plants from the harmful effects of the cold air. However, when numerous farmers in the region are all pumping groundwater at the same time, there can be problems with sinkhole formation. Indeed, Tihansky notes in her work that January is particularly problematic for farmers and those impacted by sinkholes resulting from pumping of the aquifer.

Terrance Bengtsson (1987) discusses this problem in some detail and mentions an area of Hillsborough County where these problems were common. Specifically, he defines the Dover area between Plant City and Tampa as a zone where sinkholes can occur due to cold season pumping. The Dover area has a geologic setting similar to the rest of west-central Florida. A surficial aquifer that consists largely of sands and sandy clays of Holocene-Pleistocene age exists there, and beneath this unit are the Bone Valley Formation and underlying Hawthorne Formation. This unit serves as an aquiclude between the overlying sandy layers and the limestone layers, beneath which are the Floridan Aquifer. Directly beneath the Hawthorne Formation is a thick sequence of limestone. The highest unit here is the Tampa Formation. In order from top to bottom, the Suwannee Formation, Ocala Formation, and Avon Park Formation underlie this formation. This thick limestone package contains the limestone beds that host the waters of the Floridan Aquifer.

Bengtsson reports that the USGS has been keeping records of water levels using a series of wells in the Dover area since 1958. These wells indicate that there have been tremendous pressures on the aquifer during extreme freeze events in the winter, particularly in 1977, 1979, 1981, 1983, and 1985. The stresses occurred because huge amounts of water were pumped to prevent freeze damage on the 4,000 acres of strawberry fields, 1,700 acres of citrus groves, and 300 acres of landscape plant nurseries in the area as of 1985. During the freezes of 1985, the water withdrawals peaked at 3,064 gallons per day per acre. Since then, numerous other freezes have impacted water withdrawals in the region.

Bengtsson (1987) reports that during the 1985 freeze, due to the lowered water table, there were approximately 350 complaints regarding dry wells from residential homeowners in the Dover area during the cold spell in January. This is not particularly surprising, because of the 3,200 known wells in the area, 85 percent were for domestic supply. The pumping effectively lowered the water table below the well points of the homeowners.

In addition, during a two-week period in January and February 1995, 27 sinkholes formed in the area. Most of these formed during a 12-hour period during the freeze, when water withdrawal was most rapid. However, several formed during the two weeks following the initial formation of the sinkhole swarm. Bengtsson believes that there are two factors that influenced the formation of sinkholes during this event. First of all, the withdrawal of water lowered the water table and reduced buoyancy for the bedrock overlying void spaces that would normally be filled with water. Second, as irrigation water infiltrated the soils to the surficial aquifer, the water mounded on top of the limestone to add weight to the already stressed limestone resting on top of the pumped voids. Since most of the pumping was done during cool temperatures at night, little of the water was lost through evapotranspiration. Also, there are very few surface streams in the area to carry water out of the fields, so most irrigation water infiltrated to the subsurface. Once in the surficial aquifer, it mounded above the relatively impermeable Hawthorne Formation.

The research of Bengtsson (1987) and Tihansky (1999) follows a study by Metcalfe and Hall (1984), who discuss a freeze event that occurred over five consecutive nights in the Dover area in January 1977. During this period, groundwater was pumped to a point that the potentiometeric surface dropped as much as 18 meters. In this particular case, the pumping was done to protect the region's important strawberry crop. During the cold spell, air temperatures reached a low of -4 degrees Celsius. As in the later event described by Bengtsson, numerous local residential wells went dry or had other problems associated with the drawdown. Although there were 22 reported sinkholes during the event, the authors speculate that a much greater number may have formed. They believe that many went unreported, as they may have occurred in fallow fields or vacant land and were not discovered until long after the notoriety of the event subsided.

The sinkholes varied in size from 1 to 6 meters wide and 0.5 to 3 meters deep. There was some property damage associated with the formation of the sinkhole swarm. There was severe structural damage to one home from a sinkhole collapse and a road was damaged during the formation of another. In another case, a truck driving down a road fell into a sinkhole as it formed and its frame was damaged. In addition, a telephone pole was unearthed. There was also damage to some of the land and equipment involved in agricultural production. The berm of a fish pond collapsed, citrus trees were destroyed, strawberry fields caved, and damage to chicken cages occurred. This study, as well as those by Bengtsson (1987) and Tihansky (1999), demonstrates that we do not always learn our lessons from the past.

Given the need for freeze protection for crops during extreme cold weather events, it would be useful to examine some alternatives to the massive pumping that to this day takes place in the Dover region. Water in the Tampa Bay area is becoming scarce; the region has grown from a small town in the early 1900s to one of the largest metropolitan regions in the country. Indeed, the area has looked to a variety of alternative water sources for domestic supply. A huge investment in desalination and in the formation of a local reservoir has met with some improvements in regional water production. Also, as residential land use intrudes on the traditional agricultural zones in the region, there is a great deal of competition for groundwater resources. If farmers growing specialized crops like strawberries are to succeed in the region, they will have to do so without interfering with residents' ability to maintain a constant flow of water in their wells. If sinkholes form in a residential area as a result of groundwater pumping for agricultural purposes, there will be discord between homeowners and farmers. This situation highlights the need for the development of a good land-use plan for the region that includes the understanding of the karst landscape and wise water management.

Patton and Klein (1989) describe an interesting situation in which extensive pumping of the Floridan Aquifer caused changes not only to the karst system but also to a stream system. Their work in the upper Peace River region of Polk County highlights how natural karstic systems can change significantly when stressed by human activity. They found that in the twentieth century, a particular segment of the Peace River began to change as wells were used to extract significant amounts of water. The first notable change was the decline of the potentiometric surface and the loss in 1950 of any flow from Kissengen Springs. The pumping caused a regional loss of 15 meters of the potentiometric surface. As one would expect, there were associated losses of discharge of the Peace River. The stream no longer received artesian base flow, leaving the majority of the flow to come from overland flow. The loss of base flow effectively lowered the low-water stage, although the high-water stage was not significantly impacted as that stage was largely dependent upon surface flow during extreme storm events.

During this period, a series of sinkholes opened in the floodplain of the Peace River. The authors note Holszchuh's (1972) estimate that between 150 and 200 major sinkholes formed in this area between 1949 and 1972 and that half of them would not have formed had it not been for the pumping of the aquifer. In 1980, Patton and Klein conducted a field survey to locate existing sinkholes and karst features in the floodplain of the Peace River in their study area. They found 90 probable karst features, most of them in the high-water channel of the Peace River, although certainly several are within the low-water channel. Visits by the authors to these sinkholes during periods when the river reached out of its banks to the locations of the low- and high-water channels indicate that the Peace River lost a great deal of water through the depressions, evidenced by the fact that they observed inflow into the aquifer at all sites visited.

The loss of water was quantified by the authors and demonstrates that the overall hydrologic regime of the Peace River has changed due to the combined effects of the lowering of the water table and the formation of new sinkholes during the period of drawdown. Interestingly, the loss of discharge caused geomorphic changes within the Peace River floodplain. The lateral extent of the high-water channel is reduced and the stream has abandoned its scarped channel and carved a new terrace scarp that is much reduced from the former terrace scarp. The situation described by Patton and Klein is probably similar to others in Florida where the regional water table has been lowered.

Tihansky also notes some sinkholes induced by other activities. For example, she states that it is likely that spray-effluent irrigation caused a swarm of sinkholes to develop from the excessive loading of water on the surface. One way to get rid of sewage effluent is to spray the effluent or deposit the sludge on land surfaces where it can go through a natural cleansing process. The amount of effluent sprayed at the particular location discussed by Tihansky was equivalent to 7.4 meters of rain (Trommer 1992). As Florida receives roughly between 1.3 and 1.5 meters of rain a year, the 7.4 meter total was a huge volume of water for the landscape to absorb. The weight of the water on the land surface likely caused the failure to occur.

Tihansky also recounts how a massive sinkhole formed beneath a gypsum stack in the phosphate region of west-central Florida. Gypsum stacks are created as a byproduct of phosphate mining in the region. The region holds one of the richest phosphate reserves in the world. Mining of the phosphate ore is done through strip mining, with the overburden used to re-create the landscape after the buried ore is removed. When the ore is processed, it is separated into usable materials and waste materials. One of the waste materials is gypsum. Unfortunately, the gypsum is slightly radioactive and there are very few uses for the massive amount of material that is produced annually.

Thus the gypsum is piled in huge masses called stacks. These stacks are dominant topographic features in the Tampa Bay area because they rise hundreds of feet above the land surface. Tihansky reports that a large sinkhole, with a diameter of 160 feet and a depth of 400 feet below the highest elevation of the stack, opened beneath one particular gypsum stack in 1994. A huge volume of contaminated water and sediment entered the water table when the sinkhole collapsed (figure 3.7).

It is quite likely that the imposed weight of the gypsum stack induced the formation of this spectacular feature. When one considers that the depth to bedrock is only about 60 meters, the massive weight of the gypsum stack and associated moisture must have greatly stressed the bedrock beneath the feature. In addition, the gypsum slurry is acidic and can aggressively dissolve limestone. Thus the weight, in conjunction with enhanced solution, likely worked together to cause this environmental disaster. In order to prevent a similar failure, the phosphate industry now assesses the susceptibility of new

Figure 3.7. The famous gypsum stack sinkhole collapse in New Wales, Florida. Photograph © Scott Wheeler/SILVER IMAGE Photo Agency.

gypsum stacks to subsidence by locating underground voids prior to a stack's being built. When underground voids are found, they can be plugged with concrete. In addition, the new gypsum stacks must be lined to prevent leaching of the acid-rich water.

Currin and Barfus (1989) examined the distribution and characteristics of sinkholes in Pasco County, discovering that between 1970 and 1988, 168 sinkholes formed. Like Tihansky, they note that sediment cover is an important factor in the distribution of the location of the sinkholes. Most of the sinkholes were located in the western half of the county, where sediment thickness is approximately 0–100 feet. Although sinkholes can form any month, they find a distinct seasonal trend to sinkhole formation. Most occurred between March and May, the pronounced dry season, and between July and August, the pronounced wet season. It is interesting to postulate that the occurrences are likely affected by annual variations. Extreme wet and dry years are likely times of elevated sinkhole formation as well. They find that, on average, 24 sinkholes form in Pasco County each year.

The authors also provide morphometric information about the recently formed sinkholes. They find the average sinkhole dimensions to be 10.23 feet long, 9.15 feet wide, and 9.15 feet deep (I keep the English units to maintain accuracy). The vast majority of sinkholes studied were less than 10 feet deep. Only a few were over 30 feet deep. Likewise, most sinkholes were less than 15 feet wide or long, with only a few wider or longer than 40 feet. Examining sinkhole density is helpful in predicting future risks from sinkholes. The authors note that sinkhole densities ranged from 0 to 25 sinkholes per 4 square miles over the time period studied, with the greatest density present in the western half of the county. Some higher density areas do exist in the eastern half of the county on the Brooksville Ridge and in the Tsala Apopka plain, but the authors speculate that the reason for the difference in density is that the surface cover is much thinner in the western portion of the county. That is not entirely accurate, however, as portions of the Brooksville Ridge have very limited cover on top of bedrock. In addition, it must be noted that the western portions of Pasco County may be undergoing enhanced solution due to the fact that the area is underlain by a mixing zone environment where freshwater and saltwater come together to create a more aggressive environment for the solution of limestone.

Clearly the formation of sinkholes in the Tampa area causes problems for people. Later I discuss some of the policy issues associated with modern sinkholes. However, geologically, the sinkholes in the Tampa swarm provide an opportunity to examine the causes of sinkhole formation and the meaning of their distribution. While sinkholes can form anywhere in a limestone terrain, it is evident from Tihansky's work that they occur most commonly in areas with a thin sedimentary cover. However, their geographic distribution is clearly linear and they thus follow the traces of particular structural and lithological elements in the subsurface. Indeed, work done in the region on joint patterns by Upchurch and Littlefield (1988) shows a rough north-south trending joint pattern. This is the exact same pattern seen in the sinkhole patterns in the Tampa region. Certainly other factors can influence this linear pattern, such as the presence of paleo-shorelines, but it is evident that geological structure and lithology greatly influence the pattern of sinkhole expression on the landscape. Klimchouk and Lowe (2002) and Klimchouk (2005) discuss how karst settings and speleogenesis can also influence the distribution of sinkholes in places like west-central Florida.

It is also clear that many of the sinkholes in the area are uvalas, or coalescing sinkholes. For example, the lakes in the Carrolwood area of Tampa are bordered by multiple curves, whereas many other sinkholes clearly formed from a single event as they left nearly precise circular forms on the landscape.

Bloomberg et al. (1987) characterize four sinkholes on the campus of the University of South Florida in order to assess the nature of the lithologies in which they form and to understand sinkhole sedimentation. Two of the sinkholes are clearly visible on the landscape. Two others were not visible at the time of study: one was filled, and the other was reactivated while researchers conducted tests for the project. Cone penetration tests and standard penetration tests were conducted to assess the subsurface geology of the sinkholes. Bloomberg's group found that the stratigraphy of the area consisted of five basic units, listed here from the surface to the base of the limestone with thicknesses noted: (1) a mottled brown sand (3.2 meters); (2) a mottled red silty sand (2.9 meters); (3) a grayish green clayey silt (1.9 meters); (4) a brownish yellow clayey silt (2.0 meters); and (5) limestone.

The researchers divided these five units into three main stratigraphic packages. The first package consists of units 1 and 2. This combined unit is interpreted to be part of a single marine transgressive depositional event. Units 3 and 4, which separate the transgressive package from the limestone, are also considered a single stratigraphic package. The authors note that there are two possibilities for the formation of this clayey silt package: the material is marine sedimentation associated with the Hawthorne Formation, or the package consists of residuum left from the downward solution of limestone.

They note that Carr and Alverson (1959) believe that it took 1.5 to 3.0 meters of limestone to produce 0.1 meters of residual clay, and they suggest that since residual chert is found in the silty clay package, it is likely residuum. The final stratigraphic package is the underlying limestone beneath the sediments.

The researchers also examined the conduit fill of these sinkholes. They state that the material within the solution pipes is largely noncohesive sediment. Specifically, they note that sinkhole failure results in long quartz sand–filled pipes. This information is important because the pipes serve as a breach of the aquaclude between the surficial aquifer and the Floridan Aquifer. Thus in places where these pipes exist, water from the surficial aquifer can easily travel through the porous and permeable noncohesive sand-filled pipes into the Floridan Aquifer. Sinkholes, whether they are filled or not, can therefore pose an environmental threat to the quality of the main source for drinking water in Florida.

The characteristics of the subsurface in the region as described by Bloomberg et al. (1987) on the campus of the University of South Florida are similar to those described by Gilboy (1987) in the Interwellfield Project Area in southern Pasco County. Here, ground-penetrating radar, followed by coring, was used to describe the nature of the subsurface and to identify paleo-sinkholes that might have formed in the area. Gilboy (1987) notes several interesting things about the subsurface that help to explain the history of karst in the region. He finds that although the land surface is now flat, the topographic expression of the limestone bedrock is irregular. Gilboy attributes the irregularity to the presence of a karst terrain prior to the deposition of the sediments in the region. Thus the area clearly had well-developed karst landforms on the surface prior to the Quaternary marine transgression, which deposited a blanket of sand on top of the area.

In addition, Gilboy notes that the clays on top of the limestone, which some have theorized are marine deposits, actually undulate with the topography of the limestone. The clay does not thicken at low spots or pinch out at high spots. This clearly points to the nonmarine origin of the clays. If they were marine in origin and formed as part of a marine transgression after the formation of the karst landforms, they would have filled the low areas prior to being deposited on top of the high areas of the undulating bedrock. Instead, the clays rest on top of the bedrock in near uniform thickness regardless of topographic location. If the clays were deposited in a marine environment prior to the development of the karst features, one would expect some degree of lithification of the clays or that some of the clays would have eroded from the higher areas for deposition in low areas prior to the Quaternary transgression. Thus some of the clays in the Tampa area are most likely a residuum that rests on top of a limestone surface. As solution has occurred in the limestone from the surface downward, the clays are left behind as impurities.

This conclusion has an interesting implication for the stratigraphy of Florida. Because these residual clays are in the same stratigraphic setting as the Hawthorne Formation, perhaps a component of the Hawthorne is some portion of residual limestone. There is no question that the Hawthorne Formation in Florida is a complex depositional unit. However, might the lowest unit of clays at the surface of the limestone in some areas where the Hawthorne is present be residuum? If that is the case, perhaps more emphasis needs to be placed upon puzzling out the nature of the contact between the limestone and the surface sediments in Florida.

Several lakes in the Tampa area have very interesting hydrologic characteristics due to the karstic nature of the subsurface. Stewart (1987) describes several of these lakes. Lake Tarpon in Pinellas County, for example, once had rapid changes in water chemistry due to its connection, through sinkholes, with the ocean. In fact, Lake Tarpon went through flushing cycles where saltwater that flowed into the lake was later flushed out again. Specifically, water entered a sinkhole in a bayou approximately 5 kilometers from the lake and flowed southeast underground to a sinkhole outlet near the lake when lake levels were low. This caused the salinity of the lake to increase. When water levels increased through natural runoff, water flowed in the opposite direction, causing the lake to drain seaward through the sinkhole connection. In order to stabilize this situation, the sinkhole was separated from the lake in 1969 using a dike system. This greatly reduced the salinity in the lake.

Stewart (1987) also describes how the subsurface karst deleteriously impacted Lake Grady, a reservoir impounded by an earthen dam, in the southwestern portion of Hillsborough County. Two years after construction, a sinkhole formed beneath the lake, causing it to drain $2.6 \times 10^5 m^3$ of water. The sinkhole that formed was quite large—5 meters deep, 14 meters long, and 12 meters wide. The stream that fed the reservoir now flowed directly into the sinkhole and disappeared underground. Unfortunately, the sinkhole drained directly into the local aquifer and had a negative impact on the well water of nearby residents who had wells in the Floridan Aquifer at depths of 18–35 meters. Water samples from wells revealed that after the collapse, the color of the water changed, and alga, bacteria, and other organisms were present. To solve the problem, a berm was built around the sinkhole and the sinkhole was plugged to prevent water from entering. This appears to have solved the problem. However, the berm was breached when the lake overflowed its banks a year later, and the plug sank a meter. When this occurred, water quality in local wells again deteriorated. The berm was repaired, eliminating problems as of Stewart's report (1987). This study is interesting as it so clearly demonstrates how quickly water can flow from the surface into the subsurface aquifer.

In addition to the two lakes described above, Stewart also examines the hydrology of several other sinkholes. He provides figures on the tremendous amount of water that flows through the sinkholes into the subsurface during normal and extreme flow events. In addition, he describes how some lakes in the region go through extreme annual high and low water levels due to flows into and out of the Floridan Aquifer. These extreme high and low water levels are not in the range of natural seasonal fluctuations seen in other lakes. Thus it is clear that some lakes in the Tampa region are much more connected to the Floridan Aquifer than are other lakes, thereby demonstrating the complex, and largely unknown, nature of the subsurface.

One of the most comprehensive studies of sinkholes in Hillsborough County is by Upchurch and Littlefield (1987), who examined the distribution of ancient and modern sinkholes in 12 USGS 7.5-minute topographic quadrangles in a variety of different land-use and geologic settings in the county. The quadrangles included Sulphur Springs in central Tampa; Citrus Park and Odessa in northwest Hillsborough County; Wesley Chapel, Zephyrhills, Thonotosassa, and Plant City West in the north-central portion of the county; and Brandon, Dover, River View, and Lithia in the south-central area of the county. In order to map the location of ancient sinkholes, the researchers used aerial photos and satellite images to discern the extent of these depressions. Modern sinkholes were mapped using data obtained in the field as well as from the Florida Sinkhole Research Institute and the U.S. Geological Survey.

A total of 2,303 ancient sinkholes were identified in the study area. They are located in three main sinkhole terrains defined by the absence or presence and thickness of sediment cover on top of the bedrock. One area is defined as having no or thin cover, another is defined as having cover, and the third is defined as having thick cover. The authors note that the distribution of ancient sinkholes is highest in the bare or thin cover areas, where the land area covered by sinkholes is approximately 10 percent. In areas of covered karst, the extent of sinkholes ranges depending upon thickness of the cover. In places where it is thin, sinkholes cover 7 percent of the land surface. In areas where the cover is thickest, sinkholes cover only 1 percent of the land area. The sinkholes located in the third area, the thickly covered karst, are rare and account for only 1 percent of the land area.

Upchurch and Littlefield (1987), when they mapped the distribution of the sinkholes, found that there were particular regions of extensive ancient sinkhole formation and areas where there is very little evidence of ancient sinkhole formation. The greatest area of ancient sinkhole formation occurred in the Sulphur Springs karst terrain and the Brandon karst terrain. Both of these areas are situated where the karst is bare or thinly covered. The areas with the fewest ancient sinkholes are in the southern portion of the study area, where sedimentary cover over bedrock is quite thick, and in the Green Swamp wetland region in the northeastern portion of the study area.

A total of 179 modern sinkholes were reported between 1964 and 1985 in the study area, which averages to 7.6 per year. However, the researchers state that sinkhole formation did not occur at a steady state. There were times when no sinkholes formed and there were times when there was above average sinkhole formation. The maximum number of sinkholes that formed in any year in the area was 29. Most of the sinkholes occurred in the two areas where there was no or thin cover or in areas of covered karst. There were two localized areas of sinkhole formation identified: the Sulphur Springs area, which coincides with localized ancient sinkhole formation, and the Dover area. The Dover area, studied by Bengtsson (1987), was discussed earlier in this chapter. Here sinkholes were induced from excess pumping of water for agricultural applications during freeze events.

A significant point of the authors' study is that mapping ancient and modern sinkholes can be of great assistance in developing a sinkhole risk model. Indeed, they show that there is some correspondence between ancient and modern sinkholes and that sinkhole formation can generally be predicted as likely to occur within some particular higher risk areas. This idea is similar to the risk mapping done to predict floodplain risk in stream systems. Mapping of particular frequency of flooding risk influences the type and availability of insurance to property owners in flood-prone areas.

It is interesting to speculate on our ability to produce risk maps for sinkholes. Certainly there are areas where sinkholes are more likely to occur, but sinkholes can occur in any area of limestone terrain in Florida regardless of the thickness of the cover. There may also be a problem in using topographic and visual evidence from aerial photos and satellite imagery to map ancient sinkholes. As noted earlier in this chapter and as will be noted in discussions of sinkholes in other parts of the state, many ancient karst features are covered with sediment and can be reinduced through natural or manmade processes. While it is likely that modern sinkholes will occur in areas where ancient sinkholes once formed, there are numerous examples of sinkholes forming in locations where there is scant topographic or visual evidence of ancient sinkhole formation. Thus I believe that while topographic maps and satellite and aerial photos are useful, they are not of significant value in predicting sinkhole occurrence. As we become more familiar with our subsurface through geophysical techniques, we are likely to improve our ability to assess locations of sinkhole risk.

One of the areas where there is a great deal of modern sinkhole formation in the Tampa Bay area is in Pinellas County. Each year there are many reports of damaged structures as a result of ground instability. Interestingly, not all of this damage is caused by sinkholes alone. Pinellas County has a unique deposit of shrink-swell clays that is near the surface in a thin north-south line in the north-central portion of the county. Here, the clays react to the moisture conditions in the soils. During wet times, the clays expand, and during dry spells, the clays contract. When these changes occur, the ground can shrink or heave, causing cracked building foundations and damage that appears very much like sinkhole damage. As we will see, damage from shrink-swell clays is not covered in homeowner's insurance, whereas damage caused by sinkholes is often covered. As we can imagine, this leads to numerous legal, policy, and scientific difficulties.

In addition, there are layers of peat in Pinellas County, and elsewhere in Florida, that can cause shrinking of the land surface. When water levels are low, buried peat deposits can oxidize, causing an overall reduction in the volume of the substrate. Depressions that resemble sinkholes can form and can damage homes. Unfortunately, the typical homeowner's insurance does not cover damage from peat oxidation either. Given this situation, there has been a great deal of attention on the location of sinkholes in Pinellas County in order to better understand their distribution and the risk to property owners.

Beck and Sayed (1991) completed an examination of the sinkhole hazard in Pinellas County in order to assist with urban planning. To conduct their research, they studied recently formed sinkholes using the sinkhole database kept by the Florida Sinkhole Research Institute, sinkholes voluntarily reported by insurance companies, and sinkhole reports from government agencies. A total of 42 sinkholes were documented through these reports. The majority were cover-collapse sinkholes, with only eight subsidence sinkholes. The cover-collapse sinkholes are mainly found in the northern portion of the county between Lake Tarpon and the coast. The authors of the study note that the cover over the bedrock is thinnest in this location. However, it must be noted that cover-collapse sinkholes were found in most portions of the county. Nevertheless, the authors find that in the areas of highest concentration, a sinkhole can be expected to form once every twenty years in a one square mile area. Interestingly, the month with the greatest number of sinkhole formations is April. In fact, that month had double the number of sinkhole occurrences of any other month. January, February, June, and September also had notable numbers of sinkhole formations. In spite of this, the authors state that there is not a clear relationship between seasonality and sinkhole formation.

The study's authors also examined the length of the sinkholes that formed in Pinellas County during the period of study (the length of a sinkhole being defined as the longest axis of the sinkhole form). It is thought that in any sinkhole area, the distribution of the sinkhole length value is logarithmic. What this means is that there are a large number of very small sinkholes and very few large sinkholes. This proved to be true in Pinellas County as well. There were 15 sinkholes with a length interval of 0–10 feet (note: I keep the original English units for accuracy), 13 sinkholes with a length interval of 11–20 feet, four sinkholes with a length interval of 21–30 feet, one sinkhole with a length interval of 31–40 feet, no sinkholes with a length interval of 41–50 feet, and one sinkhole with a length interval of 51–60 feet. Clearly, the majority of sinkholes that formed were quite small. It is evident that large, catastrophic sinkhole events are quite rare.

In Brinkmann et al. (2007), the researchers note that urbanization greatly modified sinkholes throughout the region. Hundreds of sinkholes were destroyed or in other ways modified throughout Pinellas County. Littlefield (1984) examines the distribution of sinkholes in a portion of Hillsborough County and attempts to relate the distribution to photolinear features in the region. The researchers define photolinear features as aligned "sags, sinkhole lakes, drainage, or soil tonal zones" (Littlefield 1984, 190). In their examination of the region, they found that there were clear photolinear features that aligned with regional structural joint patterns in the state. They were surprised that the lineaments did not align with local features but instead were affected by the broader structural elements of the subsurface bedrock.

In order to assess the role of structural elements on karst formation in the region, Stewart and Wood (1984) conducted a detailed study of the fracture traces at two locations: the Cross Bar Ranch well field in Pasco County and a limestone quarry in Citrus County near Crystal River. At each location, they carried out a variety of geophysical experiments using resistivity and microgravity techniques. Interestingly, the lineament at the Cross Bar Ranch site where the geophysical techniques were employed is 1 kilometer long and is evident through a series of aligned depressions, whereas the lineament at the Crystal River site is only several meters long and is identified by the presence of a linear swale.

The geophysical analysis revealed that the subsurface at the two sites is very different. The subsurface at the Cross Bar Ranch lineament has a distinct V-shaped depression in the bedrock that is filled with sediment beneath the zone of lineation. In contrast, the lineament at the Crystal River site has a bedrock high beneath the surface in which is located a small V-shaped depression. This indicates that photolinear features in the state of Florida are complex and not controlled by bedrock structure alone. There are likely geomorphic factors that influence their presence.

Of course, sinkholes can be locations where pollution easily enters the Floridan Aquifer, as many of them connect the ground surface with the most important source of drinking water in the state. Nowhere is this problem of more concern than in the Tampa Bay area, where the majority of the drinking water comes from groundwater sources. Trommer (1987) studied five sinkholes in west-central Florida—Blue Sink and Peck Sink in Hernando County, Bear Sink and Hernasco Sink in Pasco County, and Curiosity Sink in Hillsborough County—in order to examine the potential for pollution of the aquifer. Some of these sinkholes are located within or on the edge of the Brooksville Ridge. Trommer also studied an internally drained area of Hillsborough County near Brandon comprising a total of six study sites. While these named sinks are the focal point of the drainage, other sinks exist within their drainage basin and are included in the overall study on pollution potential in the region.

The Blue Sink area is within the Brooksville Ridge, and the sink itself is near the ridge crest. The land use in the drainage basin is largely agricultural and forest, although the City of Brooksville is located within a portion of the Blue Sink drainage basin. Potential pollution sources within the drainage basin are major roadways, storm-water discharge from the City of Brooksville, septic tanks, and agricultural activities, including dairy farming. The Peck

Sink drainage basin is connected to and southeast of the Blue Sink drainage basin in Brooksville. While Blue Sink is located on the Brooksville Ridge, Peck Sink is located on the fringes of the ridge. The Peck Sink basin has considerable urbanization within its borders. Part of the City of Brooksville is located in the area, as are several suburban housing developments. In addition, at the time of the report (Trommer 1987), sewage effluent was sprayed on the lands in the area as a means of waste disposal. (Septic systems and roadways are potential pollution sources.) Bear Sink is located in northeastern Pasco County in a zone of coastal lowlands. Bear Sink drainage actually consists of 13 sinkholes, two of which are quite large. When flowing, Bear Creek flows into Bear Sink. The land use in the drainage basin is largely urban, as development has occurred rapidly in the Hudson area centered on U.S. 19. Some areas in the eastern portion of the Bear Sink drainage basin remained rural as of the 1987 but have since largely been transformed into residential development. Thus, the greatest risk of pollution through the sinkholes in the Bear Creek drainage basin is from storm-water runoff from residential lands and roadways. There are also hundreds of septic systems in the area that could deleteriously affect the water quality.

The other sinkhole drainage basin studied in Pasco County is Hernasco Sink, which is part of the Crews Lake drainage basin. The sink is located in a broad sinkhole plain that extends north from Tampa to the fringes of the Brooksville Ridge. Several other sinks in the vicinity of Hernasco are part of the broader Crews Lake drainage system. They can become connected when high water reaches over the banks of Crews Lake and the other sinkholes to form a large water body during extreme wet events. Land use in 1987 was generally agricultural while some forested land remained. However, there is some rural land use, particularly around Masaryktown, that discharges septic waste into the subsurface. A regional airport and U.S. Highway 41 traverse the drainage and are potential sources of contamination, as are several illegal dumps. Some of these dumps are located directly in sinkholes. Curiosity Sink is located in north Tampa near the intersection of Fowler and Florida Avenues. The land within the drainage basin of the sinkhole is entirely urban with largely residential and commercial uses. The homes in this area utilize a municipal sewage system. The area of the Brandon internally drained basin is located in central Hillsborough County. While not associated with any particular sinkhole, the entire region is internally drained. In 1987, the area was undergoing a transformation from agricultural land uses, particularly citrus growing, to residential land use. The transformation continues today, making the Brandon region one of the fastest growing areas in the state. Nevertheless, there are numerous sinkholes in the area that drain to the aquifer and surface water.

The author of the study notes significant variations in sinkhole form and hydrology among the sites, indicating that not all sinkholes within any particular region are alike. Specifically, Trommer (1987) states that three of the sinkholes—Blue Sink, Curiosity Sink, and Hernasco Sink—can be considered individual drainage points that bring water directly to the Floridan Aquifer. In contrast, Peck Sink and Bear Sink are part of a multiple-sink drainage system. It is believed that not all sinks in these two systems are connected to the aquifer. Large caverns in the Floridan Aquifer that allow tremendous volumes of water to flow underlie two of the systems, Bear Sink and Curiosity Sink. Four of the basins—Bear Sink, Curiosity Sink, Crews Lake, and the Brandon area drainage basin—have a surficial aquifer present above a clay unit (the Hawthorne Formation) that serves as an aquaclude for the Floridan Aquifer.

Trommer believes the potential for pollution to the aquifer is high in four of the basins: Curiosity Sink, the Brandon basin, Peck Sink, and Blue Sink. Curiosity Sink was judged a high risk because of the density of the land use development in the area and the proximity to major roadways, such as Fowler Avenue, Florida Avenue, and Interstate 275. In addition, the sinkhole drains directly to the aquifer. Likewise, the Brandon basin was considered to be high risk due to the rapid rate of development and the fact that there are multiple sinkholes in the basin that are connected to the aquifer. Peck Sink was evaluated as high risk because it is used as a storm-water drain for the city of Brooksville. Finally, Blue Sink is considered to be high risk because there have been reported pollution problems in the area, likely from storm-water runoff and septic systems.

Two of the sinks, Bear Sink and Hernasco Sink, were evaluated as having a moderate risk for pollution of the Floridan Aquifer. Bear Sink was judged to be moderate risk because the flow-through rate to the aquifer is rapid and because the aquifer flows toward the Gulf of Mexico near the sink, making use of the water for human consumption unlikely. Nevertheless, one must note that the pollutants, if present, could enter the Gulf of Mexico through springs or could remain in brackish aquifers near the coast. The Hernasco Sink was judged a moderate risk due to the limited development in the area in 1987. As population pressures have increased in this area, it is likely that the aquifer is at greater risk today than it was in 1987.

The research reviewed above demonstrates that humans and their landscape are linked, particularly in karst areas. The land use we impose on the earth's surface has a distinct and direct effect on earth systems. Development in the drainage basins where the sinkholes in the study are located has a direct impact on the water quality, and the water quality, in turn, can impact the type of human behavior that governments allow in a particular area. What is particularly interesting about Trommer's study is that each sinkhole drainage basin is different. They each react in different ways to hydrologic pressures and to human activity within the basins. It is clear that the surficial and subsurface geology is important to the hydrology of sinkholes. As we have seen over and over in Florida, the thickness of the unconsolidated land cover above the Hawthorne Formation is a strong factor not only in sinkhole formation but also in the hydrology of the systems. In addition, the presence of conduits is important to the rate of flow through sinkhole systems. This flow-through rate can influence the residence time that the water is exposed to harmful pollutants. Finally, the location of the sinkhole relative to other features, such as another body of water, the ocean, or a ridge, influences the hydrology and the form that the sinkhole will take.

A study by Sinclair et al. (1985) classifies the landscape in west-central Florida by the types of sinkholes that can form in different areas. Much of the Tampa area is classified as Zone 2, where the cover on top of the limestone is thin or where the limestone is barely covered. The researchers note that the area is very well drained by stream systems and that it has been covered several times by seawater during Pleistocene high-water stands. The high water caused marine and coastal sediments to fill older sinkholes. The Tampa area also has some land that is in Zone 5, particularly in the east Tampa and Brandon area, and some land that is classified as Zone 4, in the northwestern Tampa area. Zone 5 has a sand cover that is 25 to 150 feet thick. Between the sand and the limestone is a thick layer (25–200 feet) of impermeable clay. Here sinkholes of the cover-collapse and cover-subsidence varieties are most common. In Zone 4, 25–100 feet of sand and clay deposits cover the limestone. There are numerous cover-collapse sinkholes that formed in this area. Many of them have distinct orientations at 40–45 degrees and at 110 and 115 degrees.

The Orlando Sinkhole Swarm

The Orlando sinkhole swarm is geographically more extensive than the Tampa sinkhole swarm. The sinkholes also tend to be larger and spaced closer together here, and they form when groundwater levels are low (Wilson and Beck 2005). Indeed, in some areas, such as Lake County, the landscape looks like Swiss cheese (figure 3.8). As noted earlier, the sinkhole swarm extends from Lake Kissimmee in the south to the Ocala National Forest in the north and from Lake Jesup in the east to the Green Swamp in the west. As is the case with the Tampa sinkhole swarm, many of the Orlando swarm sinkholes are fringed with residential development. Indeed, the downtown area of

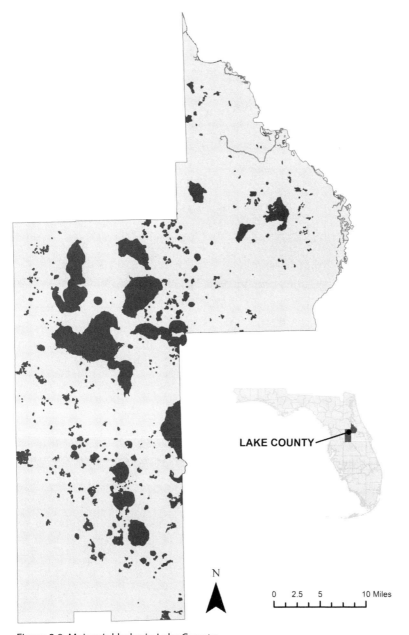

Figure 3.8. Major sinkholes in Lake County.

Orlando is built around several large sinkhole lakes, including Lake Lucerne, Lake Eola, Lake Cherokee, Lake Olive, Lake Davis, Lake of the Woods, and Lake Copland. Dozens and dozens of other named sinkhole lakes dapple the Orlando region.

Although large sinkholes seem to dominate the area, hundreds of smaller sinkholes are present as well. Schmidt (1997) notes that the extent of a sinkhole is largely dependent upon the thickness of the overlying material. The void collapse may be quite small, but due to the overlying cover, the depression spread can be quite large, like an hourglass shape. Thus the larger lakes in the Orlando district are not all that different from the smaller sinkholes that formed in areas where the cover is not as extensive. While this makes a considerable amount of sense, it must be noted that the topographic differences between some of the large lake areas and the smaller depression zones must be considered. In many cases, the larger depressions are on higher ground than the smaller sinkholes. It is certainly possible, therefore, that some of the larger lakes are older depressions and better developed than depressions in the lower areas. At some point the ridge lands were islands and the low areas were covered by seawater. Therefore, the higher elevation areas had a longer period of time for karst processes to dissolve the underlying bedrock.

Kindinger et al. (1999) discuss the geology and evolution of lakes in the Orlando area. In their work, they conducted seismic studies of more than 30 lakes in order to assess their formational history. In addition, they classified the lakes they studied into four geomorphic types: (1) active or collapsing phase, which they classified as young; (2) transitional phase, which they classified as middle age; (3) base level phase; and (4) polje. They further describe the base level phase as a time when the sinkhole is no longer active and is plugged and in the process of filling in with sediment. The polje lakes are defined as having a multicomponent formation process with multiple sinkholes creating the complex depression feature.

Their classification scheme is based on when the depression is actively forming and when the mechanics of sinkhole formation are no longer active. The active phase sinkhole lakes are defined as having steep-sided edges with a small aerial extent. As time passes, the walls slump and creep inward to create more gentle slopes. The transitional phase is defined as a time when sinkhole formation begins and the topography of the lake bottom flattens as sediment fills in low lying areas and as laminated beds are deposited. Interestingly, this period can go through spectacular fluctuations during which the sinkhole outlet into the aquifer may plug and reopen with some frequency. This phenomenon causes lakes to suddenly drain and refill as time passes. Lake Jackson in Highlands County is one such lake and will be discussed in more detail later. The base level phase is defined as a time when the lakes begin to fill significantly with sediment and the lake bottom flattens considerably. As the lakes begin to fill with sediment, wetland vegetation can begin to creep in, going through a biological succession to the point that no open water is present. The poljes in the area, as noted above, are identified as having flat floors with multiple karst features present beneath the lake surface.

Although the authors studied more than 30 lakes, they focused their effort on four central Florida lakes. The Blue Pond in Clay County is considered an active phase lake; Lake Magnolia, also in Clay County, is considered a transitional phase lake; Lake Jessup in Seminole County is considered a base level lake; and Orange Lake in Alachua County is considered a polje. The work of Kindinger et al. is an intriguing effort to classify Florida lakes using a formational history approach. Instead of examining lakes by their type of formation, they classify the lakes by their destruction history. This approach is similar to that used by William Morris Davis in his efforts to describe landscape evolution (Davis 1899). The active phase corresponds with Davis's youth phase, the transitional phase corresponds with Davis's mature phase, and the base level phase corresponds with Davis's old age phase. The polje sinks are fundamentally different from the other sinkhole lakes and must be considered separately. However, one wonders whether this idea might be taken further in an attempt to find features that have undergone the full succession of infilling and biologic succession. How many sinkholes lakes once existed on the Florida landscape that now are totally filled with sediment? Certainly there are examples in the region of sinkholes that are fully overgrown with vegetation. Cypress domes, for example, are locations where biological succession has allowed cypress trees to fill sinkhole lakes. Although Kindinger et al. focus on only a few lakes, it would be a useful exercise to develop a more thorough classification of sinkhole lakes in Florida. This could be greatly assisted by the online *Florida Atlas of Lakes* (wateratlas.org 2011), which is produced by the Florida Center for Community Design at the University of South Florida.

Thorp and Brook (1984) examined the distribution of sinkholes in a 9-by-12 kilometer area in Orlando in order to ascertain if there are any particular alignments that can indicate bedrock structural control of their distribution.

In order to do this, they conducted a double Fourier series analysis of x, y, and z coordinates of 3,780 points. From this, they were able to discern a series of waves that were oriented along the landscape.

Thorp and Brook found that there are three main trough directions—north-northeast, northeast, and north-northwest—accounting for about 40 percent of the topographic variability in the region. In addition, they found that sinkhole axes are roughly parallel to the northeast- and northwest-trending troughs. Specifically, they found that the axes of 90 percent of the sinkholes are found associated with the trough and that the larger sinkholes are found in places where the wave troughs intersect. The researchers believe that the association is related to the underlying bedrock, specifically structural characteristics of the bedrock. While this may be true, it would be interesting to add other layers to this model, such as thickness of overburden on top of the limestone, thickness of the Hawthorne Formation, and the location of other geomorphic features such as the prominent ridges that traverse the state. This is certainly possible using modern geographic information techniques. Such an effort would greatly assist in understanding the relationship between geologic structure and the pattern of sinkhole distribution in Florida. Right now, there is not a clear model that puzzles out the many possible controls on sinkhole distribution.

In a study of the sinkhole density of the Forest City Quadrangle, Bahtijarevic (1989) compared the sinkhole density of existing sinkholes to that of sinkholes that had formed recently. Forest City is located just north of Orlando and is subject to modern sinkhole formation. As of 1989, Bahtijarevic noted that the Forest City area had the highest reported incidences of reported sinkhole formation in the Orlando area. In order to assess the distributional elements of these newly formed sinkholes, she constructed digital terrain models of the quadrangle and calculated sinkhole area and density for the newly formed sinkholes and the existing sinkholes. She found a total of 385 sinkholes within the quadrangle and obtained information on reported sinkholes from the Florida Sinkhole Research Institute's database. Unfortunately, she does not list exactly how many sinkholes were reported. Nevertheless, she provides some interesting results. She found that there is no correlation between the locations where the old sinkholes formed and where the new sinkholes formed. What this means is that new sinkholes were forming in areas where there were no sinkholes in the past. This is interesting from a genetic and historical point of view. Does this mean that the sinkholes that have already formed might have formed under different environmental conditions than the recently formed sinkholes? Is the older karst some sort of paleokarst with the modern impetus for karstification (hydraulic modification driven by water withdrawals for human consumption) creating a new series of karst landforms? Or has the area where sinkholes existed prior to the study an area where the maximum sinkhole formation could have occurred? In other words, have most of the voids in the area of old sinkhole formation been played out? Perhaps the areas where new sinkhole formation occurred are zones that still have void space to allow for the formation of sinkholes. Regardless of the answers to these questions, it would be interesting to examine the causes behind the distribution of the newly formed sinkholes and the older sinkholes in this quadrangle. Perhaps there are geologic reasons for the distribution separate from the digenetic evolution of the karst. There could be variations in the thickness of the land cover or variations in the depth of the water table that are responsible for these differences.

A study by Snyder et al. (1989) examines the Crooked Lake region within the Lake Wales Ridge in order to ascertain how sinkholes connect to the underlying aquifer. The lake received its name because it consists of a series of rounded basins that are connected to each other in an apparently haphazard manner. On this lake, the researchers were able to construct a bathymetric survey using a seismic profiler. In addition, they created sub-bottom profiles using EG&G UNIBOOM and ORE GEOPULSE seismic systems. The results of the bathymetric survey revealed that the lake floor was flat in the portion of the lake that was located in the lower elevation segment of the Lake Wales Ridge and that the lake had significant topography consisting of basins and knolls in the portion of the lake that coincided with the higher reaches of the Lake Wales Ridge. Clearly the authors found that the remnant bedrock on the ridge affects the topography of the lake bottoms.

The results of the seismic scans of the sub-bottom revealed numerous filled solution pipes within the lakes, indicating that the complex lake formed as a result of numerous subsidence events. The filled solution shafts were 8–12 meters deep in lake bottom sediments and extended 20–40 meters into the Miocene siliclastic sediments on top of the Miocene carbonates where the collapse occurred. The solution pipes do not significantly impact the topographic expression of the lake bottom. This can mean that the solution pipes fill relatively quickly after the formation of the pipe, leaving little local topographic trace on the lake bottom. What is particularly fascinating about this study is that the authors provide evidence for dozens of sinkholes being responsible for the formation of Crooked Lake. Certainly many of these were quite small. Indeed, the authors found evidence for only one significant event. The remainder of the sinkholes formed from small individual events as indicated by the sub-bottom seismic scans. What also is interesting is that the authors found evidence of continued collapse on the fringes of the lakes, which leads to an interesting question regarding the evolution of lakes in this region. If the lakes formed from numerous small events, could the lakes grow along their own edges? What this would mean is that there could be lakes that are still expanding as solution continues on their fringes.

We are just starting to understand how long the sinkholes in the region have been forming. In order to assess the formation history of a particular sinkhole, Hansen et al. (2001) analyzed the pollen and conducted radiocarbon tests of organic material found within a core taken at the Peace Creek site in eastern Polk County. The Peace Creek site is an interesting location: a filled sinkhole that was exposed during strip-mining activities east of Bartow between the Lakeland Ridge and the Winter Haven Ridge. The researchers note that several of these filled sinkholes are evident on the landscape as marshy low areas in an otherwise flat landscape. The core that was collected in the filled sinkhole consisted of "interbedded peat, sand, and organic silt and clay layers" (Hansen et al. 2001, 683). The authors could not obtain a maximum depth of sediments as the depth was beyond the limits of the coring. Nevertheless, the core accounts for 71 meters of sedimentary history.

The radiocarbon dates show that the near surface had dates of 3510 BP ± 70 (0.30–0.33 meters in depth) and 6820 BP ± (0.40–0.43 meters in depth). A sample collected from 7.0–7.5 meters had dates over 30,000 years, and a sample collected from 18.9–19.1 meters had dates over 36,990 years. These last two dates are beyond the limits of radiocarbon dating and cannot be used with certainty. Regardless, it is clear that the sinkhole is filling slowly with time, accumulating a mixture of sediment and organic material through some mixture of colluvial, palustrine, and alluvial processes. The pollen assemblages help to further provide information on the formation of the sinkhole as many of the pollen samples were taken below the maximum sampling depth of the C-14 samples. Based on the pollen assemblages, the authors concluded that the sinkhole formed during the Neogene after the deposition of the Bone Valley Member of the Peace River Formation. The time of formation likely was after 2.8 million years, which is at the same time as the beginning of widespread glaciation in Europe and North America and the associated worldwide drop in sea level.

The pollen results also indicate that the sinkhole remained a wetland throughout most of its existence. This is interesting, as most believe wetlands to be temporary features on the environment that fill through alluvial, colluvial, and palustrine sedimentation with time. The authors believe that the sinkhole may have remained moist due to the sedimentation of fine particles early in its formation, which prevented significant leakage. As water tables dropped, the sinkhole maintained its wet characteristics by holding water that ran into it from its basin. During wetter periods, the wet conditions in the sinkhole were assisted by an associated rise in the regional water table. It is clear, however, that Florida's filled sinkholes hold a vast reservoir of information about the geologic history of the state and that much can be learned from studies such as that conducted by Hansen et al. (2001).

There is tremendous void space under the Orlando area (Kimrey 1978). Indeed, there are two known cavernous zones that are productive for water supply: an upper producing zone at a depth of 45–180 meters and a lower producing zone at a depth of 335–460 meters. In the past, storm water was drained directly into the upper producing zone in order to alleviate flooding and manage sheet flow in the area. Like many karst zones of the world, the Orlando area has few surface streams, and sheet flow flooding, particularly during the high rainfall summer months, is a problem. Currently, the preferred practice is to store water in water-retention ponds.

Due to the high growth rate of the Orlando area, there is a great deal of concern about sinkholes due to pollution potential and because of their ability to store and transmit storm water. In a 1996 study of the Wolf Branch sinkhole basin in Lake County, Schiffer (1996) notes that there is the potential for contamination of the aquifer system through the Wolf Branch Sinkhole, particularly during extreme hydrologic events. In her study, she found through dye tracing that the sinkhole was not directly connected to the upper surficial aquifer but was connected directly to the Floridan Aquifer in some way. Thus concern exists in this area that sinkholes provide direct links into the subsurface drinking water source for the region.

In a 1989 study, Ryan notes that there is the potential for leakage from central Florida lakes in the Orlando district regardless of whether they are plugged or not. He completed a numerical model using field data collected from an observation network installed around a small lake in the Orlando area. He specifically found that loss of water from the lake was not dependent upon the presence of fractures or other kartsic features. Instead, he notes, the leakage is dependent upon the stagnation point or "the point of minimum head along the potentiometric divide separating the lake from underlying ground water" (Ryan 1989, 43, 45). What this means is that while leakage along fractures or raveling zones may be important in rapidly transmitting water from sinkholes directly to the aquifer, any sinkhole lake, regardless of the presence or absence of fractures or raveling zones, has the ability to transmit water directly into the aquifer system. This information is important in understanding that depressions in Florida all have the potential to leak pollutants into Florida's important aquifer system.

Certainly there is a risk to residents in the Orlando area from sinkhole formation. Perhaps the best-known sinkhole in the state, the Winter Park Sinkhole, frightened many Floridians and others living in karst terrain by the reality of what can happen when sinkholes occur suddenly (the Winter Park Sinkhole will be discussed in detail in an upcoming chapter). However, less dramatic sinkholes occur regularly and are part of the ordinary passing of time in Florida. Indeed, we have become accustomed to seeing reports of sinkholes on local television news and we regularly are exposed to advertisements for companies who specialize in sinkhole law or in the purchase of sinkhole-damaged homes. Thus it is understandable that a great deal of engineering science has gone into the areas of land and structure stabilization. For example, Goehring and Sayed (1989) recount a case history of a sinkhole stabilization project in Orange County near Orlando. In their study area, a sinkhole 2.4 meters deep and 9 meters wide opened suddenly adjacent to a roadway and beneath a 137-cm effluent line. While engineers were filling the sink with cement, the sinkhole dropped again, creating a depression that was 15 meters wide and that raveled to a depth of 38 meters.

The situation left the engineers with a difficult situation. The 137-cm concrete effluent line was left standing unsupported through a span of 15 meters and failure appeared certain. This situation required quick action to stabilize the pipe. The sinkhole was filled immediately with sandy soil, effectively reducing the risk of failure to the pipe. Then, cement grout was pumped to depths of 6–38 meters to fill the void space underground and to seal the raveling zone. The combined efforts of filling the depression and pumping cement grouting into the void stabilized the land surface and prevented costly damage to an important effluent pipe. This is an example of common occurrences throughout Florida's karst terrain. Projects that require special engineering skills help to stabilize roads, buildings, and land surfaces.

Kuhns et al. (1987) review a specific case study in Maitland in which a large office building was constructed over an active karst area. This project, undertaken in northwestern Orange County and southwestern Seminole County, was in an area that has long been known to be a zone of active sinkhole formation, and therefore there was concern for the stability of the foundation of the large, four-story office complex and associated parking garage that was to be built on the site. The authors found several risk factors to sinkhole formation, noting that sinkholes in central Florida are more likely to form when the following conditions are met (Kuhns et al. 1987, 367):

1. Thin confining beds
2. Large difference in elevation between the water table and the Floridan Aquifer potentiometric level
3. Limestone in relatively close proximity to the ground surface
4. Effective aquifer recharge
5. Fracture zones within the limestone

These conditions exist in the Maitland area where the case study took place. Thus it was important to evaluate the subsurface conditions in order to assess the specific problems that could be encountered from ground instability during the life of the office building.

Several methods were used to evaluate the subsurface. Standard test borings, piezocone soundings, ground-penetrating radar, and seismic refraction helped assess the presence of karstic features in the subsurface. Of course, the presence of topographic depressions was also investigated to help explain the existing karst landforms and their relationship with subsurface features. A total of three depressions were found at the site upon topographic investigation.

The standard test borings revealed that the area consists of three main units. The uppermost is a layer of loose sand that is common throughout much of central Florida and is described by many as a marine sand of Quaternary age. Underlying this is a zone of loose sediment that is a mixture of "intermixed very loose clayey sand and very soft clay and silt" (Kuhns et al. 1987, 369). This loose zone is common in the Maitland area and has been encountered by others doing subsurface borings in the region. In fact, it is likely that it is a localized stratigraphic phenomenon. The authors note that the material has fluid properties and that it reacts on drilling almost like an underground void. Beneath this zone is limestone covered by residual clay.

The loose zone is the area that is of most concern. It is thought that many of the structural problems in the area are caused by collapse of the overlying sands into the loose zone. This could be brought on by sinkhole action.

However, the researchers note that a simple model of raveling of sands into the subsurface to form sinkholes does not fit the field observations. Interestingly, the thickness of the loose zone varies across the study site and is dependent upon the depth to the limestone stratigraphic package. Where the depth to limestone is thin, the loose zone is thin. Where the depth to limestone is thick, the loose zone is thick. This leads to some intriguing ideas about the formation of the loose zone. If the thickness were uniform, it could be speculated that the zone might occur as an in situ weathering feature. However, since the material thins at the topographic highs of the bedrock, it is likely that the material is depositional and its thickness influenced by post-depositional erosion. Indeed, the material may contain paleosols that formed prior to the deposition of the overlying sands.

Another interesting feature was found in the borehole investigations. The authors probed the center of one of the depressions in the area and found a 6.7-meter-thick deposit of muck on top of silt and sand. These data indicate that the rate of organic deposition, at least in recent decades, was greater than the rate of sedimentation. What this means is that sheet flow across the area does not bring a sizable volume of sediment to the depression. Perhaps, due to the rapid rate of soil infiltration, sheet flow is minimal, and infiltration directly through the sands into the subsurface dominates the area's hydrology.

The piezocone soundings, a form of resistivity, helped to identify the nature of the loose zone at the Kuhns et al. study area. Ground-penetrating radar was of limited use because the researchers were able to obtain a record to a depth of only 3.1 meters. Nevertheless, they were able to discern that the sediments beneath the sinkhole described above were very different from the surrounding sediments. This helped to delineate the extent of the muck deposit within the depression. The seismic refraction survey was of no use to the researchers due to the near-surface presence of a shallow groundwater table and due to background interference from construction and roadway activities.

The Kuhns et al. (1987) study revealed that a stabilization program was required prior to the construction of the office building. This was accomplished by pumping grout (cement) into the loose zone to strengthen the layer, a common practice in the region that has served to effectively reduce the risk of foundation failures from ground instability.

The Maitland area has many sinkholes present on the land surface, and it is evident that sinkholes existed in the study area. However, the greatest threat to the construction of the building was the presence of the loose zone, consisting of unconsolidated sediments. What is the link between this loose zone and sinkhole formation? Are the sinkholes somehow not karstic in nature but instead caused by failure of the sediments on top of the loose zone? Does the loose zone liquefy under particular conditions, and is flow dependent upon the overlying pressures? Regardless, the distribution of the loose zone is certainly dependent upon the buried karst features in the Maitland region. In addition, there is ample evidence for the formation of sinkholes in the area caused by failure above a limestone void. However, the loose zone described by Kuhns et al. (1987) demonstrates the variability of the subsurface geology in Florida.

There is also a risk to the region from activities that discharge pollutants onto the land surface. For example, Stangland and Kuo (1987) discuss the potential for pollution to the aquifer system from the rapid-rate land application of sewage. They describe a variety of land-application systems used in Florida and how the rapid system allows for the application of up to several centimeters of processed sewage onto the land surface. Of course, this is only permitted if the area is absent of any karstic features that could connect with the subsurface. Stangland and Kuo (1987) describe a case study where Seminole County hoped to develop a rapid application system on two parcels of land. Neither parcel had any evidence of sinkhole activity, but ground-penetrating radar was used to try to assess whether or not there were any relict sinkholes in the area that could connect the sewage effluent to the aquifer. Hildebrand and Oros (1987) also document the search for a land application site in Osceola County to handle the excess effluent produced in the city of Kissimmee. After a significant amount of time in site investigation, a suitable site devoid of relict sinkholes was found. The location required that sewage be pumped fourteen miles from the city.

The results of the ground-penetrating radar work proved that there were filled relict sinkholes in one of the field sites. By taking cores in the area of the paleo-sink, the authors were able to delineate appropriate areas where land application would be appropriate.

The next chapter details sinkhole formation and distribution in the more rural areas of the state.

4

Sinkholes in the Ridges and in North and South Florida

After reviewing the geologic setting of sinkholes across the state with a discussion of urban sinkholes near Tampa and Orlando in the previous chapter, I turn here to some of the other sinkhole regions in the state, particularly sinkholes on the ridges and in north and south Florida.

Sinkholes of the Brooksville and Ocala Ridges

Numerous sinkholes exist on the Brooksville Ridge and the Ocala Ridge. However, they are not clearly present on satellite or aerial photos as many of them are not filled with water. Neither are they wetlands. Instead, many of the sinkholes on these ridges are well drained. Thus one must use topographic maps or LIDAR mapping to fully appreciate the density of sinkholes in these areas. The Brooksville and Ocala Ridges are two of the largest ridges in a ridge complex that parallels the Florida peninsula. A map showing the main ridges on the Florida peninsula can be seen in figure 3.5. There is some disagreement about the formation of the ridges. Nonetheless, the ridges are distinct topographic regions in the state. Indeed, it is evident that they were topographic highs for long periods of time, and portions of them were islands during Quaternary high-water stands (Barco 2001).

Interestingly, the ridges are made up of bedrock that is older than the bedrock in the surrounding lowlands. The Ocala Ridge is made up of Ocala Limestone, and the Brooksville Ridge is made up of Suwannee Limestone and Ocala Limestone of Oligocene and Eocene age. Surrounding the ridges are Miocene and Pleistocene rocks and sediments. Thus the ridges were likely islands during higher stages of sea level during particular times in the Cenozoic. Significantly, the area surrounding the ridges is mantled with tens of feet of Quaternary marine sand, whereas the ridges are capped with a thin cover of marine sand. While some of the sand may have eroded off the ridges, it is likely that the ridges were never zones of extensive marine sedimentation due to their elevation. While certainly underwater for some periods of high sea level, they were definitely above water for a much greater period of time than the surrounding landscape.

The importance of this fact is that the surfaces of the ridges have undergone karstification for a greater period of time than the surrounding landscape. Karst processes require a steady supply of freshwater and are significantly inhibited under saline conditions. Thus the ridges are places where karstification has been going on since the early Cenozoic, and much of the karst landforms present on the land surface today are among the oldest karst features in the state of Florida. It is not surprising that the karst landforms on the ridges are complex.

Karst features on the Brooksville and Ocala Ridges include sinkholes, uvalas, solution valleys, disappearing streams, springs, and caverns. The sinkholes and uvalas are among the most extensive depressions in the state. The ridges themselves are complex entities and have not been fully studied to understand their complex formation or geomorphic history. Subregions within the ridges can certainly be identified. Both ridges have a distinct lack of surface streams. The drainage is multibasinal, with surface water draining through the ground directly or flowing into sinkholes where the water enters the underground aquifer systems. This is characteristic of karst landscapes and leads to significant environmental challenges as the state's population grows.

The Brooksville Ridge contains numerous coalescing sinkholes or uvalas and several karst hills that seem to be transitioning into magotes. Indeed, the landscape is transitional between a karst plain, defined as a flat plain punctuated by sinkholes (such as in Tampa), and a cockpit karst landscape, consisting of numerous magotes surrounded by a flat surface of former sinkhole bottoms. Although we certainly have karst plains in Florida in places where the landscape has recently emerged from under the sea, such as the Tampa area and the Everglades, the state does not have any true cockpit karst landforms. Cockpit karst is common in more mature karst landscapes, such as in Puerto Rico, Cuba, and China.

The Ocala Ridge also contains numerous coalescing sinkholes or uvalas. However, the distance between sinkholes is greater in the Ocala Ridge when compared with the Brooksville Ridge, so the sinkhole density is greater in the Brooksville Ridge. Why is there such a difference? This question has not been studied although it is likely that the lithology, depth to bedrock, and joint patterns of the Ocala Ridge are dominant factors in the difference. The Ocala Ridge is made up mainly of the Ocala Limestone, whereas the Brooksville Ridge is made up mainly of the Suwannee Limestone. The Ocala Limestone contains more dolomite than the Suwannee Limestone. Dolomite is slightly less soluble than calcite, the dominant mineral in the Suwannee Limestone. Perhaps this difference causes less widening of joints and thus less sinkhole formation in the Ocala Ridge. Of course, another difference could be the joint density of the bedrock. Solution is most active along joints and joint intersections. The greater the joint density, the better the opportunity for solution and associated sinkhole formation. Although there have not been any comparative studies of the two ridges, the distribution of joints within the ridges could be a significant factor in the development of sinkhole density. There also may be differences in speleology. While karst in Florida is largely viewed as forming from waters flowing from the surface downward, some scientists have suggested that the karst, at least in part, may be forming from waters flowing upward to the surface (Klimchouk 2007).

Interestingly, several large springs exist within the Ocala Ridge and flow as spring runs to the eastern fringes of the ridge. Juniper and Rainbow Springs are striking examples of Florida springs and are easily visited in the Ocala National Forest (figure 4.1). These springs discharge at sinkholes within the Ocala Ridge. Springs associated with the Brooksville Ridge, in contrast, discharge mainly on the western fringes of the landform. Here, the spring runs flow from the spring head short distances to the Gulf of Mexico. Some notable springs on the fringes of the Brooksville Ridge are Homosassa Springs, Weeki Wachee Springs, and Crystal River.

Both ridges are quarried for lime rock and are thus important to the local rural economy. Quarrying activity has revealed a tremendous amount of information about the subsurface geology of the ridges. Unfortunately, due to Florida's limited topographic expression, it is difficult to obtain data about the subsurface. Many Florida geologists long for road outcroppings or exposed mountainsides and have found creative ways to ascertain information about the subsurface geology through geophysical investigations, coring, and examination of quarry exposures when they are available. The quarries in the Brooksville and Ocala Ridges contain abundant information about the karstic nature of the subsurface. For example, as rock is removed, caves are often revealed. In addition, it is common to find widened joints containing sediment that has raveled in from the surface. As an aside, fossils found in the quarries have helped to date the age of the rocks.

Figure 4.1. One of several springs in the Ocala National Forest.

The quarrying activity in both ridges demonstrates that there is significant void space underground and thus there is still tremendous potential for sinkholes to form in these ridges. Although these ridges contain some of the oldest karst features in the state, it must be understood that karst processes continue to occur in the ridges. Certainly, solution is not taking place as rapidly as in the zone of saturation, but in these unsaturated portions of the ridge, numerous caves and voids formed during times of higher water tables, probably during times of higher sea level. A study by Florea et al. (2007) maps many of these caves in the Ocala Ridge. Similarly, Brinkmann and Reeder (1994 and 1995) and Reeder and Brinkmann (1998) describe some of the caves in the Brooksville Ridge. These authors demonstrate that although there is not extensive void formation in the unsaturated zones, there is a threat of sinkhole formation from the collapse of overlying bedrock into the caves. There are many caves in both ridges that have not been mapped. However, the caves that we know about provide ample evidence that the ridges are evolving into a more pitted landscape.

Pasco County, which is a transitional area between the sinkhole swarm in the Tampa area and the sinkholes of the ridge country, is an interesting place to examine sinkhole formation. Distinct sinkhole zones in the county have been mapped by Fretwell (1988). The easternmost portion of the county has very few sinkholes because the area is a broad wetland floodplain of the Hillsborough River and Withlacoochee River. West of this lowland is the Brooksville Ridge. Here, the ridge contains broad and deep paleo-sinkholes, and new sinkhole formation is rare. Drainage is largely internal but contains portions of the Hillsborough River basin. Overall, however, most of the drainage on the ridge flows into sinkholes or into intermittent streams leading to the Hillsborough River. West of the ridge is a lowland extending to the coast where numerous small sinkholes form with some regularity. The sediment cover in this location is relatively thin, and there have been problems with structural integrity in many neighborhoods here. This area has many intermittent streams leading to named sinkholes, including Bear Sink, Round Sink, Rocky Sink, Stratomax Sink, Briar Sink, Golfball Sink, Smokehouse Pond, Coffee Sink, Rock Sink, Hernasco Sink, Crews Lake Sink A, and Crews Lake Sink B. It is believed that these sinks connect directly to the Floridan Aquifer. The water that is not collected through sinkholes is captured by the Anclote and Pitlachascotee Rivers, which drain directly into the Gulf of Mexico. A number of springs drain to the surface in Pasco County.

Fretwell notes that a number of high-volume well fields in Pasco County, including the Cross Bar Ranch, Cypress Creek, Starkey, and the South Pasco well fields, significantly impact the surface hydrology. Surface drainage and lake levels are all impacted by pumping due to regional groundwater effects.

A study conducted in the northern section of the Ocala Ridge (Arrington and Lindquist 1987) examines the formation of large sinkholes near Interlachen, a community approximately ten miles west of Palatka and a few miles north of the northern fringe of the Ocala National Forest. Orange Springs is very near this area, and the study authors are interested in how large sinkholes formed there. They note that although karst topographic expression is clearly evident, the conditions that caused the formation are different from those in other areas. Specifically, they note that the limestone is covered by between

35 and 130 meters of sediment, that the areal extent of the sinkholes is large, and that surface karst landforms occur fully within the sediments (this sedimentary body is called the Citronelle Formation) overlying the limestone. In fact, the authors state that the underlying limestone is not exposed in any of the sinks that they studied in the Interlachen area.

Arrington and Lindquist (1987) provide a detailed description of the sinkholes in this part of Florida. One of the morphometric variables they examined is the length-width ratio of the major to minor axes of the sinkholes. They found that the ratio was 0.9, a figure that suggests that the sinkholes have a great deal of circularity. The authors believe that the circularity of the sinkholes in the region is due to the presence of thick layers of overlying sedimentary material. When the sinkholes form, the sediments filter in hourglass fashion to the subsurface to create a relatively uniform depression. The authors note that the influence of bedrock characteristics, such as joint patterns, would be minimal due to the thick cover of sediment. Thus there is limited elongation of sinkholes. The authors also note that the deeper sinkholes are generally larger in size than the smaller sinkholes. In the field, the authors had the opportunity to record observations of a sinkhole that formed in the area in 1985. They found that when the sinkhole initially opened, it had a diameter of 7 meters. They do not state the initial depth, but after three days, the sinkhole had a diameter of 75 meters and a depth of 12 meters. The expansion of the diameter, they explain, is largely due to slumping (and supposed infilling of the initial depth).

In order to assess some possible explanations for the great size of the depressions in the Interlachen area, the authors calculated the volume of material removed from the surface to the subsurface by calculating the volume of the void left by the formation of sinkholes in the region. What they found is that the volume of sediment removed is much larger than the volume of void space in the subsurface. If the karst is still active, then, they needed to come up with an explanation for where all the sediment is going and where it is likely to go in the future. In order to do this, they developed a model for sinkhole formation in the area. Their model brought sediment into limestone voids in the subsurface through a piping effect. Transportation is accomplished as sediment falls into single pipe-like openings into the void. The pipes eventually expand into the thick overlying sediments and bring sediment into the void. Thus much of the lost sediment is within piping features in bedrock and in the sediment body itself as collapses occur through this process. When the pipe eventually makes its way to the surface, the topographic expression of the deep collapse failure occurs. With time, the initial collapse slumps to form a wide, gently sloping depression. Sediment may still filter into the subsurface to cause an expansion of the depression.

This model is similar to others that have developed over the years for the formation of sinkholes in covered karst areas. What is interesting about the Arrington and Lindquist (1987) model is that they suggest that the piping is part of a "network of smaller solution channels with dimensions on the order of a few meters" (36). In other words, the sediment is not stored in some large void where a giant failure occurred beneath the sinkhole. Instead, the void space is a more complex system of voids in which sediment is transported to the subsurface through pipes for storage in large tubular voids in the subsurface. Arrington and Lindquist (1987) believe that there is sufficient volume in the bedrock in the Interlachen area to store the volume of sediment is transported in this way. They developed their model before the development of precise geophysical tools that could better assess the subsurface of the region. It would be an interesting exercise to evaluate their model in light of what can be learned through a detailed geotechnical investigation of the nature of the subsurface. Nevertheless, their model certainly is useful in explaining the genesis of the large, circular depressions in the ridges.

Although the regions of Ocala and Brooksville are developing rapidly, one of the greatest threats to the region is the pollution of the aquifer through sinkholes adjacent to roadways. Some of the worst pollutants we have in our environment are found in storm-water runoff from roadways. Interstates and other important roadways traverse the Brooksville and Ocala Ridges. In a 1993 study, Padgett examines the risk to aquifer leakage through sinkholes adjacent to Interstate 75 in Alachua County. He used aerial photos to find the location of sinkholes and to evaluate the associated risk. The problem was highlighted by a case study he discusses in which a tanker truck hauling hazardous waste overturned and exploded, leaking its waste into a closed depression system and associated wetlands. This accident demostrated the risk to the state's aquifer system and caused a rethinking of how drainage systems adjacent to interstates are built.

Sinclair et al. (1985), in a classification of areas of sinkhole types in west-central Florida, place the Brooksville Ridge area within Zone 3, defined as having a 15- to 30-meter sand cover above limestone. In addition, the authors note that the area has no surface runoff, indicating that drainage is internal. What this means is that surface water readily enters the subsurface, where it actively corrodes limestone. Piping of sand into the subsurface voids causes sinkholes to form slowly, although sudden collapse may also occur.

Sinkholes of North Florida

The sinkholes of north Florida are part of a broad sinkhole plain that extends into Georgia and associated areas of Alabama. In many ways, these sinkholes are similar to those in the Tampa area in that they are associated with the Floridan Aquifer in a broad plain. In addition, both regions contain both paleo-sinkholes and recently formed sinkholes. Active sinkhole formation is an issue in both regions.

Troester et al. (1984) compare the sinkhole depth frequency in a portion of north Florida with other sinkhole regions of the world. The area of Florida they used in their study included four USGS topographic quadrangles along the Suwannee River. Based on their description, it appears that they measured sinkholes west of the Cody Escarpment although this is not particularly clear. The researchers compared the north Florida sinkholes to several Appalachian sinkhole regions, to sinkholes in central Kentucky, to sinkholes in northern Puerto Rico, and to sinkholes in the Cervicos karst region in the Dominican Republic on the island of Hispaniola. Using available topographic maps, the researchers measured each sinkhole in the area for depth. From this data, they were able to plot the number of sinkholes of a particular depth. This exercise was done for each sinkhole region, and the results were compared. In addition, the average depth and sinkhole density was calculated for each area.

The results starkly distinguish Florida from the other regions. Florida has the greatest sinkhole density, but it also has the shallowest sinkholes of all areas studied. In addition, the researchers found that many of the Florida sinkholes are very large and are often compound sinkholes formed from multiple collapse events. Although the authors didn't mention this, the depth is probably controlled by the near surface water table. In order to assess depth, the authors used contour lines available on maps. This is problematic in that many of the sinks in Florida extend beneath the water into lakes or are filled with loose sediment that has washed off the loose, unconsolidated deposit of Quaternary sands. In addition, the authors note that in some of the other areas studied, there is a thick deposit of sediment on top of the limestone where collapses occur. This could certainly inhibit the formation of a broad distribution of sinkholes. Comparison of the whole spectrum of sinkholes across all study areas reveals that the tropical karst of the Dominican Republic and Puerto Rico have much deeper sinkholes than the Appalachian and Kentucky regions. It is interesting that Florida does not fall between these two areas. If it did, a very clean conceptual model could be developed that suggests sinkhole depth increases with tropical conditions. Instead, the sinkholes of north Florida are anomalously shallow, suggesting that other factors, such as depth to the water table, height above sea level, and the thickness of limestone, are important factors in determining the morphology of sinkholes.

As in other parts of Florida, there is concern about environmental hazards associated with the sinkholes of north Florida. For example, Price (1989) summarizes problems associated with a hazardous waste site near Live Oak, Florida. Here, a wood preservation company operated for thirty years, treating lumber with creosote and pentachlorophenol. Unfortunately, the company's waste materials were dumped into an unlined lagoon within a paleosink basin. Given the karstic nature of the region, it was feared that there might be leakage of the chemicals into the groundwater supply. Investigations show that creosote sludge had formed at the bottom of the lagoon but that there was not any contamination of local wells. In order to better puzzle out the fate of the contaminants, the author placed several other wells in the area and collected split spoon core samples to determine where the waste was going in the region. He found that there was some migration of contaminants, but he postulated that due to the karstic nature of the landscape, the material did not flow out of the lagoon but rather into and through desiccation cracks that formed when the lagoon was dry.

This study highlights the fact that sinkhole systems are extremely heterogeneous and difficult to predict. Indeed, in his study, Price points out the importance of sound initial geologic investigations prior to interpreting the data one finds in any study.

Snyder et al. (1989) studied a portion of the St. Johns River basin near Palatka in order to determine the nature of the buried Paleogene limestone surface. The northeast portion of Florida is underlain by a thick sequence of Miocene sediments and rocks that constitute the Hawthorne Group, and beneath this is Paleogene carbonate strata. The authors explored the topographic nature of these Paleogene sediments in order to ascertain the karstic nature of the bedrock beneath the Miocene and Quaternary sediments. What they found is that the St. Johns River basin is underlain by a series of solution valleys. The overlying sediments have deformed to conform to the topography of the valleys. Interestingly, it is likely that the solution valleys formed hundreds of thousands of years ago. Thus the authors believe that the St. Johns River did not cut its valley with time but flowed as a captured stream within the ancient solution valley. As sedimentation occurred through time, the river continued to hold its valley form.

Rupert (1991) examined cave floor sediments from the Wakulla Spring in Wakulla County, Florida, in order to try to puzzle out the history of its formation (figure 4.2). Wakulla Spring discharges about 15 miles south of Tallahassee and flows in a spring run called the Wakulla River. It flows roughly southeast to Apalachee Bay, where it discharges into the Gulf of Mexico. The spring is a first-magnitude spring and discharges water from the Floridan Aquifer. Rupert believes that the spring occurred because of a post-Miocene collapse of a sinkhole.

Figure 4.2. Aerial photo of Wakulla Springs. The spring run here flows east from the sinkhole spring. Photo courtesy of Cindy Shaw.

There have been numerous explorations of the spring and the associated cave by divers. The spring pool is known to be about 100 × 200 meters, and the spring itself flows from an extensive cave system that has been the subject of a great deal of interest due to the presence of Paleo-Indian artifacts, vertebrate bone, and charred wood. One of the most interesting questions related to the cave is whether or not it was exposed at some lower sea level at a time when Native Americans and Pleistocene fauna could have inhabited the cave.

Rupert reports that a group of cave divers conducted extensive mapping of the Wakulla Cave system and found that the main cave was quite large by Florida standards, measuring 18 meters high and 36 meters long. They also found that after 270 meters, the cave splits into four caverns, the longest of which is over 1,280 meters long. During the exploration, the divers recovered five shallow sediment cores from the cave bottom in order to conduct sedimentological and palynological investigations. The maximum core length they were able to attain was 79 centimeters, although the bedrock was not reached in some of the cores.

The results of the analysis indicate that the sediment within the cave floor consists of layers of quartz sand and layers of claylike calcilutite. The quartz layers contained freshwater gastropod fragments and terrestrial and aquatic plant remains. The calcilutite layers contained diatom tests and decomposed and unidentified plant remains.

The pollen results of Rupert's tests provide very interesting evidence of the geologic history of the site. One of the most dominant pollen assemblages found there is the Chenopodiaceae, glassworts and seablites that thrive in a salt marsh environment. This is an interesting finding as Wakulla Springs is not near a sea marsh environment at the present time. Rupert speculates that the pollen entered the spring depression during an earlier geologic time, when sea levels were higher. Indeed, the author strongly suggests that the Chenopodiaceae existed in the area during periods of higher sea level and could correspond to one of two marine transgressions that occurred during the Pleistocene. The more likely of the two is the Pamlico marine transgression, which created the Pamlico Terrace throughout much of Florida. The Pamlico high-water stand occurred roughly between 130,000 and 85,000 years ago (Osmond et al. 1970). The high-water mark of the Pamlico transgression would have allowed seawater to move up the Wakulla River to the Wakulla Spring and even north of the spring. This event certainly would have allowed the formation of a salt marsh in the vicinity. Rupert suggests that the Silver Bluff marine transgression, which occurred 4,500 years ago (Tanner et al. 1989), could also have been responsible for the presence of Chenopodiaceae. However, the extent of this marine transgression caused the sea to rise only 5 feet above sea level. The limited change of the Silver Bluff transgression did have an impact on Wakulla Spring. Saltwater was able to intrude up the Wakulla River to near the present-day spring. However, the influence of the rise in sea level was not nearly as extensive as that associated with the Pamlico transgression because the saline water during the Silver Bluff transgression remained largely within the channel of the Wakulla River. Thus it is unlikely that extensive salt marsh formation occurred during the Silver Bluff high-water stand.

Rupert's study provides some interesting environmental information about the Wakulla Spring, Wakulla Sinkhole, and Wakulla Cave. Clearly the sinkhole formed prior to the formation of salt marshes in the area during the late Pleistocene. There are no salt marshes currently in the area around Wakulla Spring, and the presence of Chenopodiaceae indicates that the sink formed during a time when salt marshes occupied the vicinity. In addition, the evidence of transgression and regression occurring while the spring was present indicates that a mixing zone environment moved as sea level changed. Mixing environments are places where enhanced solution occurs when brackish water and freshwater mix and cause corrosive conditions. Solution rates in these mixing environments are greater than under freshwater conditions, and thus it is believed that the extensive cave systems often associated with coastal karst are a product of the mixing zone. With frequent documented changes of sea level in the Pleistocene, it is likely that the rates of solution changed parallel to the shoreline. It is also important to recognize that the spring provides evidence of the great antiquity of the landforms. Wakulla Springs continues to be of interest, and there have been interesting exploratory efforts conducted by the United States Deep Diving Team in the area (see http://www.usdct.org/index.php/).

Jensen (1987) makes a case for the existence of some large poljes in the north Florida area, particularly at Lake Jackson and Lake Iamonia (figure 4.3). Poljes are broad, flat areas in karst terrain that flood during high water tables. They are similar to solution valleys and sinkholes, but sinkholes are typically much smaller while poljes typically form from multiple sinkhole events that create the broad depression. Also, solution valleys are typically more linear and can have an outlet, whereas poljes are internally drained features that are not necessarily linear. Jensen makes the case that Lake Jackson and Lake Iamonia are polje features because they can have dry phases when the lakes drain. Both of these lakes are located north of Tallahassee in Leon County. Although there is abundant documentation for these occurrences, perhaps it would be better to locate the lakes within the transitional phase of development defined by Kindinger et al. (1999). As discussed earlier in the section on the sinkholes of Orlando, Kindinger et al. (1999) define the transitional phase of sinkhole lakes as one during which the lakes may experience continued sinkhole growth but, at the same time, as one where the lake can also undergo plugging and unplugging. This causes regular fluctuations of water levels as the lake fills (during periods when the sinkholes beneath the lake are plugged) and as it drains (during periods when the sinkholes are unplugged).

Smith and Randazzo (1987) report on some interesting induced sinkholes in Alachua County. Lakes are often constructed as part of urbanization in Florida. Sometimes they are constructed for drainage purposes, and sometimes they are constructed for aesthetic reasons. Typically, the lakes are lined with clays or plastic. When the lakes are filled, the weight of the water puts a stress on the lake bottom to the point that sinkholes can be induced. Smith and Randazzo describe the specific causes of ground failure at four such lakes

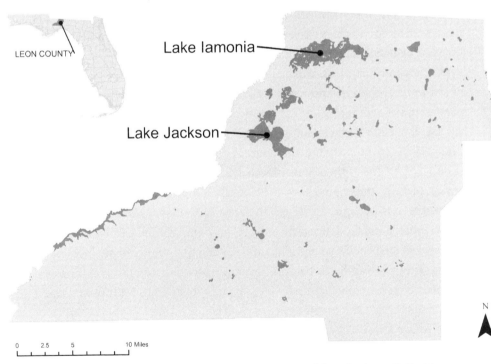

Figure 4.3. Lake Jackson and Lake Iamonia in Leon County, possible poljes in north Florida.

in Alachua County. One lake showed sinkhole formation due to two factors: (1) a localized drop in water table and the associated cone of depression caused by excess pumping to fill the lake and (2) the associated addition of the water weight onto the land surface. This situation is very similar to what occurs when rapid pumping takes place in the Tampa area to protect crops from freezing. In another lake, a plastic liner failed when a sinkhole 3 meters in diameter formed in the lake. In another constructed lake, drainage occurred suddenly and did not leave any topographic expressions on the land surface. A resistivity survey revealed that there were voids present beneath the constructed lake at a depth of 30–50 feet. These studies of induced sinkholes demonstrate that as we construct a variety of hydrologic features in our landscape, from wetlands and water-retention ponds to lakes and drainage ways, the subsurface karst system can be thrown out of equilibrium with resulting sinkhole events.

A huge volume of material is dropped into the underlying bedrock when sinkholes occur in the region. Hollingshead (1984) mapped a single sinkhole in western Alachua County, Teagues Sinkhole, in order to create a detailed topographic map from which volume measurements could be calculated. The map shows a very complex form near the center of the sinkhole, where depression contours are highly crenulated. Near the edges of the sinkhole, the depression contours become more circular. The irregularity could be caused by multiple sinkhole events or flow channels into the sinkhole that drain the surrounding landscape. It appears that at least in some locations, channels are clearly fluvial in origin. This poses some interesting questions regarding the evolution of sinkhole form. Many people imagine sinkholes to be single events that produce clear, rounded depressions on the landscape. It is evident that slope processes can act upon the sinks. Indeed, Hollingshead notes that some of the channels leading into the sinkhole are structurally controlled. His calculated volume for the sink is 189,778 cubic meters. While the sinkhole has certainly filled some since its formation, this is a tremendous amount of material to be absorbed by the subsurface. This magnitude suggests—across all of the sinkholes in Florida—the vast quantities of rock and sediment impacted through karst processes.

Another study in north Florida by Price (1984) describes the sinkholes in Suwannee and Madison Counties. Price notes that the sinkhole distributions there are largely controlled by the thickness of the surface cover of the Hawthorne Formation and the presence of a phosphatic member. He also notes that the rock in the areas with the greatest topographic expression of sinkhole formation is highly friable in the uppermost meters. This zone of soft rock is an area where the water table fluctuates significantly. Price believes that as continued weathering occurs, the sinkholes will widen and cause a general lowering of the water table to a base level. In addition, he found that many of the sinkholes in the counties are aligned with major joint patterns in the state.

In another study conducted in the same region, Upchurch and Lawrence (1984) examine the interrelationships of groundwater, geology, and karst landscape near Lake City in Suwannee and Columbia Counties. In that area, the Cody Escarpment is retreating to the northeast and trends northwest–southeast. The karst landforms to the west of the escarpment are distinctly different from those to the east of the escarpment. West of the escarpment, the area is a lowland zone containing numerous lakes. The surface sediments are thought to be residuum left from the retreating escarpment. East of the escarpment, the landscape is underlain by the Hawthorne Formation, which itself is underlain by the Suwannee and Ocala Limestones.

Upchurch and Lawrence describe three main chemical environments in the groundwater that significantly influence the solution of limestone. The first of these environments is found in the confined portions of the Floridan Aquifer. Here, the water is saturated with calcium, and thus the waters are thought to be at equilibrium with respect to the surrounding bedrock. These waters are not causing significant solution of the bedrock. This condition dominates the highlands area northeast of the scarp line. The second environment is in the recharge zones in the escarpment. Here, organic-rich water filtering from the surface into the voids in the escarpment is supersaturated with calcium and phosphorus. Thus the calcium is bound in organo-phosphate complexes, causing the water to be aggressive in solution of limestone. The water in areas where the aquifer is unconfined to the southwest of the escarpment is also undersaturated with calcium and thus aggressive in continued solution of limestone.

These results are interesting in that they provide a chemical reason for the lack of significant sinkhole development in the highland areas. These are largely the same areas Price (1984) defines as having limited sinkhole development due to the thickness of sediment cover. It is likely that both factors help to limit the development of new sinkholes. In addition, the sinkholes to the southwest of the escarpment are numerous and greater in size. Many of these are paleo-sinkholes that formed as the escarpment retreated. The current groundwater regime helps to reactivate these sinkholes or create new ones in the region. Tremendous volumes of water recycle through this system to allow new, small sinkholes to develop in the area. However, due to the fact that calcium is bound by organo-phosphate complexes in the escarpment zone, the water in these areas is the most aggressive, and it is here that one can expect the greatest development of new vertical karst features.

There are environmental problems associated with sinkholes and other karst problems in north Florida. A study by Hoenstine et al. (1987) describes problems associated with a landfill built directly on a karstic surface in Madison County, just north of the city of Madison. The authors found clear evidence of a paleo-sinkhole under the landfill when they found a sand-filled pipe within limestone that penetrated the limestone to a depth of 33 meters. The piping could have gone deeper, but at that point the maximum design depth of the drilling device was reached. Other evidence of the karstic nature of the area that was found during drilling of the wells included clay-filled cavities above bedrock and the presence of voids within the limestone. The area around Madison also has a multibasinal drainage system with numerous sinkhole depressions.

Over the years, the landfill collected household waste as well as local industrial and agricultural waste byproducts. Indeed, the authors report that metal drums containing a variety of pollutants were dumped in the landfill for many years. Unfortunately, water percolating through the landfill collects some of these contaminants and takes them to the underlying karst system. Once there, the water can enter through joints and sinkhole pipes directly into the aquifer system. Some of the pollutants that have been identified in groundwater in the area include trichloroethene, methylene chloride, and trans-1,2-dichloroethene. This example demonstrates how concentrated pollutants can cause significant regional groundwater pollution problems in a karst environment.

Another interesting area in north Florida is the Big Bend region of the state. This region, so named because it is a big bend in the shoreline of the Gulf of Mexico, extends from Citrus County on the south to Wakulla County in the north (figure 4.4). The coastline is underlain by bedrock, and thus the region has a variety of interesting coastal karst features. The landward slope of the Big Bend area is very low, so salt marsh environments extend inland tens of miles. Several runs cut through this lowland as waters from higher areas of the ridge country in the central portion of the state flow westward. Arthur (1991) reports that several springs feed the Suwannee River in Lafayette County (Alan Mill Pond, Blue Spring, Convict Spring, Fletcher Spring,

88   *Florida Sinkholes: Science and Policy*

Mearson Spring, Owens Spring, Perry Spring, Ruth Spring, Troy Spring, and Turtle Spring) and that two springs, Iron Spring and Steinhatchee Spring, drain into the Steinhatchee River. These springs have their headwaters at sinkholes.

Other notable springs flow from the ridge fringes to the edges of the coastal karst. Most notably, Homosassa Springs and Crystal Springs (and its associated spring run, Crystal River) are found here. The ground stability in the area of Crystal River has been of interest to engineers because a large nuclear power plant is located at the point where the river discharges into the Gulf

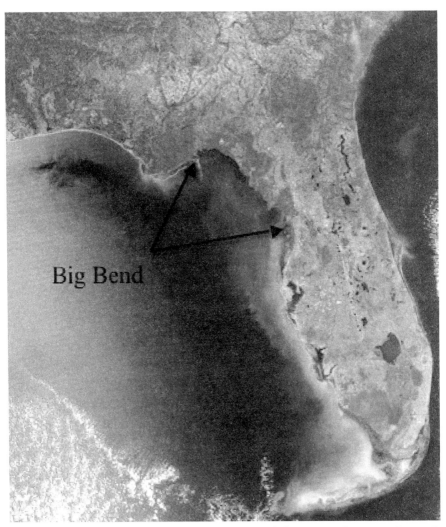

Figure 4.4. The Big Bend region of Florida, which extends from Citrus County on the south to Wakulla County in the north.

of Mexico. In fact, a study by Almaleh et al. (1993) describes several issues associated with expanding cooling towers and intake and discharge tunnels approximately 20 feet below the water table. These researchers detail how the plant is built largely on top of bedrock that has been significantly modified by karst processes. There is evidence of solution chimneys at the surface. Subsurface investigations revealed that voids exist within the surficial Inlis Limestone and the underlying Avon Park Limestone. These voids provided challenges for engineers who were designing the very large half-mile-long string of cooling towers at the site. In fact, during construction, many karst features that required specialized handling were encountered.

South of Crystal River, Homosassa Springs is a local tourist attraction due to the fact that for generations manatees have frequented the springs during cold months. The state has turned the springs into a state park and wildlife refuge, and the spring pool of Homosassa Springs is used to house injured manatees that cannot be returned to the wild.

Moving to the western Panhandle of Florida, Schmidt and Scott (1984) note that there are many depressions near Pensacola that appear to be karstic sinkholes but, in fact, formed from other processes. Here, the Citronelle Formation, which consists of a variety of sediments, including sands and gravels and massive silty clays, is at the surface. It is believed that the depressions in the region formed due to the transformation of kaolinite to gibbsite via chemical weathering. When this occurs, there is a net loss of 35 percent of the volume of the mineral. This volume loss is believed to be the cause of the depressions in the area. However, limestone is present beneath the Citronelle Formation, and some depressions in the area are likely karstic sinkholes.

Sinkholes of South Florida

While south Florida is not considered a high-risk zone, sinkholes do occur there. The entire south Florida region has limestone present near the surface. Sinkholes are also a source of concern in the Florida Keys. A 1995 study by Benson et al. summarizes the researchers' evaluation of a portion of Key Largo for sinkhole potential in an area where a 7,500-foot-long bridge was being built. There is evidence of past karstification in Key Largo, and studies by Land et al. (1995) and Land and Paull (2000) identify sinkholes in the Florida Straits. Lineated lakes and depressions were found in these areas; boreholes drilled in preparation for the bridge indicated subsurface voids; and there is a paleo-sinkhole 600 meters in diameter present in the area.

Indeed, five circular lakes are aligned on a northeast-to-southwest trending line. The bridge was to traverse Lake Surprise, one of these circular lakes, and there was concern about the stability of the ground where the bridge was going to be built. In order to assess the risk to the structure, Benson et al. broke down their work into four phases. In Phase I, they completed reconnaissance and evaluated the available literature, including conducting marine seismic profiling through Lake Surprise and a microgravity survey over Lake Surprise and beyond. Through this effort, they found a subsurface anomaly on the landward portion of the bridge beyond the lake. In Phase II, they conducted additional microgravity tests on the anomaly and conducted sub-bottom profiling. This phase helped them better define the karstic anomalies. In Phase III, the final phase, they completed detailed investigations of the shallow and deep conditions. In the shallow effort, they examined conditions to a depth of 60 meters by conducting detailed sub-bottom profiling and by drilling borings and conducting geophysical logs within the borings. This effort helped to identify joint patterns within the rocks and to interpret geological conditions. The investigation of the deep conditions, which extended to 300 meters, consisted of deep reflection data to try to identify voids underground. This effort helped to identify a large cavity system at a depth of 150–180 meters.

The Benson et al. research is interesting in that very few karst features are described in this area of Florida. Indeed, the authors note that the lakes that they identify as sinkhole lakes in the study area are often thought to have formed from mangrove growth and evolution and not from karst processes. They also point out that most karst features in the area are old due to the fact that the rock is saturated, for the most part, with brackish or saline waters that inhibit karst formation. Nevertheless, there is ample evidence in this study that there are subsurface voids in the region that could form sinkholes. The authors report that Shinn (Shinn et al. 1996) found a filled sinkhole 600 meters in diameter five miles from Lake Surprise that contains sediments to a depth of 54 meters. The researchers speculate that the reason we don't see more sinkholes in this portion of south Florida is because they are quite old and likely have filled with sediment. Thus they are not easily found on south Florida's flat landscape. However, there is evidence that the Biscayne Aquifer has extensive solution conduits (Manda and Gross 2006).

Perhaps the most important karst system in south Florida is the Florida Everglades. This is a vast system of wetlands termed the "river of grass" by Marjory Stoneman Douglas (1947). Prior to late nineteenth- and twentieth-century modification, the Everglades drained Lake Okeechobee and other basins through sheet flow across a wide, flat wetland plain. Much of the water drained into Florida Bay around the Ten Thousand Islands region of southwest Florida. This natural system extended from the coastal ridges of eastern Florida near present-day Miami and Fort Lauderdale, north to Lake Okeechobee, west to the coastal ridge lands of Fort Myers and Naples, and south to the tip of the peninsula. While much of the Everglades were wet, there were areas that would dry seasonally.

As many of the karst features in south Florida and the Everglades are subtle and greatly impacted by human modification, it is important to provide a brief summary of the modifications done to the Everglades. Through an understanding of the drainage modifications, we can see how the karst landscape was modified and how that landscape influences the modern landscape and restoration efforts. The Everglades ecosystem was one of the most productive ecosystems on the planet. Prior to human modification, the Everglades were home to a diverse community of plants and animals, including a large wading bird population. This rich environment was seen by many in the nineteenth century as being of great potential for agricultural development. Indeed, there is ample evidence of the impact of Native Americans on the natural landscape that predates Western exploration and settlement in the region (Milanich 1994). The Everglades were utilized as a rich source of food for a wide variety of individuals, and there is evidence of some Native American modification of the drainage system prior to any efforts by modern Americans. However, it was the Seminole Wars that brought Americans into contact with the Everglades. The conflict, for the first time, caused Americans to closely examine the region thought to be a vast wasteland. With the end of the Seminole Wars, and with the U.S. Swampland Act of 1850, the region was ripe for exploration and development. The act encouraged the development and drainage of the Everglades for economic purposes, particularly agriculture, and effectively transferred the region, which was owned by the U.S. government, to the state of Florida for improvements.

Due to the poor economy and the Civil War, the state of Florida did not proceed with any major efforts for a number of years. In fact, the state did not have the funds for any specific project and had to sell a significant portion of land (4 million acres) to a developer by the name of Hamilton Disston in 1881. With his efforts, the first major modifications to the Everglades began in 1881 (Light and Dineen 1994) with the connection of the Caloosahatchee River to Lake Okeechobee, the connection of particular lakes to promote drainage, and the development of the Kissimmee River as a navigable stream.

He also developed several thousands of acres for sugar cane and other crops. Given that Disston committed suicide in 1883 in response to financial ruin, it is amazing that so much was accomplished in such a short time. However, the overall economic depression that began in 1883 curtailed any further significant development of the Everglades for two decades.

In 1905, everything changed. At the time, there was intense political pressure in the state to drain the Everglades and develop the land for agriculture and to support growing coastal communities such as Miami and Fort Myers. This was an era of community boosterism, which was sweeping the United States as a whole, not just Florida. Communities were using new technologies—electricity and the telephone—and advances in engineering and building construction to showcase the products of the human imagination and spirit. Grand engineering schemes came to fruition, and the land was seen as a resource upon which the human imprint could be stamped. Thus it is no surprise that a regional grand plan for drainage and development of the Everglades came about during this golden period of American expansion. The visionary for the development of the Everglades during this era was an individual with a grand name for a grand plan—Napoleon Bonaparte Broward.

Broward was elected governor of Florida in 1905, running largely on a ticket that focused on the drainage of the Everglades. This was a popular platform driven not only by the "can do" spirit of the era but also because of disastrous floods that devastated the region's developed land in 1903. Leaving many of the day-to-day responsibilities of the governor's office to his staff, Broward oversaw much of the Everglades engineering effort that took place during this time. The work focused around an area designated by Broward as the Everglades Drainage District. Here, four major canals—the West Palm Beach Canal, Hillsboro Canal, North New River Canal, and Miami Canal—were cut, draining the region from Lake Okeechobee to the Atlantic Coast. Extending from 42 to 85 miles, the canals lowered the water table in the region to allow for the development of agriculture. Several other canals were built in the next few decades, and the Tamiami Trail, a roadway connecting the East and West coasts through the Everglades at Miami and Naples, was constructed. Each of these major construction feats significantly altered the hydrology and encouraged settlement in the region.

Disaster struck the region in 1926 and 1928 in the form of hurricanes. Each of them caused considerable damage, although the 1928 hurricane caused the most extensive damage and loss of lives. The devastation is recounted in Zora Neale Hurston's *Their Eyes Were Watching God* (1937). During the storm, water was blown out of the north and south ends of Lake Okeechobee. Tens of square miles were flooded, and hundreds of people drowned. The terrible events prodded the Everglades' engineers to focus their attention not so much on drainage for agricultural purposes as on protecting farmers and residents from future catastrophic flooding events. It was during this period that the levee surrounding Lake Okeechobee was designed and built, and drainage canals to prevent flooding were constructed. After these protections against flooding were built and the drainage improved, agricultural activities increased dramatically in the region.

Yet in 1947 and 1948, the region experienced extensive devastating flooding from two hurricanes and an annual precipitation of 108 inches (Light and Dineen 1994). Much of the southern portion of the state was under water for half the year. These problems were brought to the attention of the Army Corps of Engineers. This organization, full of talent and looking for new focus after World War II, developed a major plan for the drainage of the region in order to better manage the drainage systems currently in place and in order to create new drainage structures to better handle extreme rainfall events. Their efforts caused the water table in the region to drop 4–5 feet and effectively eliminated problems of widespread flooding. Their work was so effective that it helped to push agriculture into new areas and enhanced the ability of portions of the eastern Everglades to undergo rapid urbanization, particularly in the Miami area. The Army Corps of Engineers focused their efforts on continued development of levee and canal systems as well as the development of large water-holding areas and pumping systems. These efforts greatly changed the hydrology of the region, eliminating most natural flow.

In the 1960s ecologists and other scientists and citizens concerned about the Everglades system and Everglades National Park noticed the significant changes to the region, particularly the deteriorating ecosystem quality and the loss of natural water flow in the park. Of course, the leading advocate for the Everglades during this era was Marjorie Stoneman Douglas, who wrote the landmark work *The Everglades: River of Grass* (1947) about the Everglades system and the negative impact of man upon the land. Her work reflected a growing awareness of the impact of America's widespread efforts of land improvement upon the environment. Indeed, the vast changes to the Everglades became an important symbol to the broader environmental movement of the 1960s and 1970s.

While there was growing concern about the changes to the natural functioning of the Everglades drainage, there was also concern about the pollution running into the Everglades from agricultural fields. In 1962, Rachel Carson published *Silent Spring*, a book recounting her research on the deleterious impact of pollutants on the songbird population in North America. This work ushered in an era of research on the impact of pollution on ecosystems. With the largest population of wading birds and alligators in North America, and with a diverse ecosystem that is home to diverse animals including the Florida panther and the manatee, pollution in the Everglades, particularly agricultural runoff, became a major concern. In addition, the Everglades were seen as a source of drinking water for the burgeoning coastal populations of south Florida.

A vast number of federal, state, and local environmental rules came into effect from the 1960s to the present, including many that affected the Everglades. The Clean Water Act and the Clean Air Act certainly had a positive impact on the region. However, specific rules on the Everglades provided agreements on everything from minimum flow into the Everglades National Park to rules affecting nonpoint pollution (pollution originating from multiple sources such as storm water off of streets). The problems with the Everglades system were highlighted during 1983, an El Niño year that brought unusually heavy amounts of precipitation to the Everglades drainage basin. Excess water was diverted from agricultural lands into the natural portions of the Everglades, severely harming the ecosystem through the unnatural addition of excess water. Since that time, most of the engineering efforts in the Everglades have been focused on restoring as much of the natural environment as possible while maintaining developed lands. As one can imagine, this is a difficult task. On the one hand, one certainly would like to see the Everglades region restored and the ecology improved. However, there is a tension inherent in this effort as the south Florida region is one of the fastest developing areas in the country and there is a tremendous demand for land and water. Nevertheless, efforts toward reclamation are moving forward.

The push for reclamation of the Everglades, while certainly advanced since the 1960s, is not a particularly new initiative. As early as 1912, critics of the Everglades engineering projects spoke out against the drainage efforts (National Research Council 2003). However, modern efforts to rebuild the Everglades environment began with the synergistic publication of *For the Future of Florida: Repair the Everglades* (Marshall 1982) and Florida Governor Bob Graham's Save Our Everglades effort in 1982 and 1983 respectively. Since the 1980s a great deal of research has been undertaken on the

Everglades in order to understand how best to accomplish restoration of the system. Eventually, a document to restore the Everglades was produced by the Army Corps of Engineers in 1999 (U.S. Army Corps of Engineers and South Florida Water Management District 1999; National Research Council 2003) and passed by Congress in 2000. The law, called the Comprehensive Everglades Restoration Plan, authorized the development of the projects outlined in the plan within forty years. The projects required new engineering efforts to develop greater water storage, improve water quality, bring natural flow of water to rivers and estuaries, reconnect the natural system through the removal of levees and canals, and continue studies to improve the system. Modifications to the plan were made in 2005. A variety of projects have been completed and are underway.

Although the Federal Everglades Restoration Plan is a major effort, there are other projects underway that are improving the Everglades system. Perhaps the most visible is the restoration of the Kissimmee River. During the initial engineering efforts designed to drain the Everglades, the Kissimmee River, a widely meandering stream, was straightened to allow more rapid drainage and ease of navigation. Currently, the Army Corps of Engineers is putting back the meanders and restoring the floodplain. In addition, several storm-water treatment wetlands have been built to treat water flowing from the agricultural area around Lake Okeechobee to the Everglades in order to reduce nutrients in the storm water.

Nevertheless, much needs to be done in order to understand the Everglades system. In fact, the Critical Ecosystems Studies Initiative Science Subgroup (U.S. Geological Survey Standing Scientific Group 1996) identified seventeen areas where there are important gaps in Everglades restoration science:

1. Flexible and sustained resources are essential to an effective, comprehensive restoration effort. Some critical activities needed at early stages in the restoration process are being neglected for lack of directed resources.
2. A region-wide ecosystems approach to monitoring, support studies, and modeling in a coordinated interagency framework is the only means to attain restoration, but its achievement requires special effort and application of personnel and supporting resources.
3. Critical linkages between sub-regions are not being adequately addressed. Issues of agency authority are at times barriers to focusing efforts at problem sources.

4. Information exchange is a problem, because there is so much information in the hands of myriad sources, including local governments.
5. Monitoring projects by various agencies have not been coordinated or integrated in the restoration efforts.
6. Hydrologic models that currently exist or are under development do not have the geographic coverage required to meet region-wide ecosystem management needs or to provide the hydrological information for regional ecological models. Existing hydrological models do not extend to the coast and therefore cannot show how physical and ecological processes in the mangrove zone are affected by water management strategies. Such models are needed to support regional ecological models of wading birds and fish and to provide input to hydrodynamic models of coastal waters.
7. The most suitable current hydrological models cannot be used to test alternatives for the interagency restoration effort on a timely basis.
8. Systems of nested models are needed, in which finer resolution can be provided to address some questions and coarser resolutions can be provided to address others.
9. Modeling is not well integrated with present scientific studies, and funds for modeling usually do not include sufficient funds for special supporting studies, including verifications.
10. An objective process is needed to evaluate existing models within the context they are being used and to ensure necessary improvements are made.
11. Certain key species (e.g., apple snails) or communities (e.g., periphyton) that might be suitable ecological indicators are so poorly studied that they cannot be used as indicators. Furthermore, lack of knowledge about the response of these species or communities to hydrological and nutrient variables may seriously handicap the restoration effort.
12. Phosphorus-dowsing studies to examine the effect of loading on ecological balance in higher plant and algal communities need to be augmented by gradient studies and process-oriented studies of nutrient cycling through soils, plants, algae, and the water column.
13. While monitoring for contaminants is extensive, little interpretation of monitoring results is occurring.

14. Both tangible and intangible connections between natural and human systems need to be quantified and widely communicated while reinstatement of a sustainable system is still possible.
15. A scientifically based analysis is needed to demonstrate alternative futures under various land and water configurations.
16. Potential opportunities need to be explored for configurations of land and water that lead to ecosystem restoration and enhanced quality of life and economic sustainability in human communities.
17. There is no coordinated science program to support reduction in agricultural or urban pesticide usage. (National Research Council 2003, 26–27)

It is significant to note that some of the key areas listed above indicate a general lack of understanding of the Everglades hydrology and of how the Everglades work as a natural system. In order to fully understand the Everglades system and its modifications, one must have a grasp of the underlying geology and the karst system within the bedrock. Although much is made of the natural sheet flow that was disrupted during the construction of the Everglades drainage systems, a significant amount of water flows in the karstic aquifers underlying Douglas's river of grass.

The best way to understand the nature of the geology underlying the Everglades is to picture the region as a long-standing basin. Indeed, the basin, termed the Okeechobee basin, consists of a very thick sequence of sediment, sedimentary rock, and peat that is thicker than any other sediment of similar age in the state. Compared with the rest of the state, the thickness of sediment cover is quite striking. In north and central Florida, the thickness of post-Paleocene bedrock varies from about 460 meters in north Florida to 1,220 meters in south Florida in the Everglades. Thus, south Florida is a depositional basin that experienced enormous sedimentation for most of the middle and late Cenozoic. There is evidence that the south Florida area experienced down warping to the southeast during the Miocene (Scott 1997), probably caused by isostatic pressures from the weight of the overlying sediments.

During the Miocene, when phosphorus-rich sediments were deposited within the variable siliciclastic Hawthorne Formation, south Florida received a vast quantity of sediment. Indeed, Scott (1997) reports that the base of the Hawthorne Formation is 53 to 91 meters below sea level in the area and that the top of the Hawthorne is 10 to 15 meters below sea level. It is interesting to compare this with the depth to the Hawthorne Formation in central Florida, where it is found at or near the surface in many locations and is only a few tens of feet thick. Clearly the south Florida region was a longstanding depositional environment. During the Pliocene, thick carbonate sediments were deposited on top of the siliciclastic Hawthorne Formation. In fact, Petuch (1992) describes the region as a large lagoon that was surrounded by reefs and barrier islands during the Pliocene. This lagoon served as a rich environment for the growth of marine invertebrates, and their exoskeletons and fecal matter provided the building blocks for the limestone that was deposited. Petuch (1992) notes that the region did undergo intermittent periods of exposure during low sea level stands. It could be that Lake Okeechobee is a remnant of this lagoon. However, deposition of sediment surrounding Lake Okeechobee and sea-level changes made the lake a freshwater system in the Quaternary.

The fact that the south Florida region was largely under saline marine conditions for most of the Cenozoic is important in differentiating the south Florida karst system from that of the rest of the state. In central and north Florida (with the exception of the areas around the Jacksonville basin and the Big Bend area), parts of the state were exposed during lower sea level stands. When this exposure occurred, corrosive freshwater infiltrated the limestone bedrock, replacing the noncorrosive brackish water. During the periods of freshwater infiltration, solution occurred within the limestone to produce voids. The process was exacerbated in a mixing zone where freshwater came into contact with brackish water. Thus, in areas where the state was exposed to freshwater, extensive karstification occurred to produce the paleokarst that we see today in places such as in the Brooksville and Ocala Ridges. However, karst processes did not impact places that were under marine conditions, such as the south Florida region.

It would be interesting to investigate the extent, if any, of karstification during the brief periods of exposure of the South Florida Basin during the Pliocene. Petuch (1992) reports that duricrusts formed in the limestone during periods of exposure. However, there is little reported evidence of extensive karstification during the Pliocene.

During the Pleistocene, a large blanket of siliciclastic sediment was deposited during extreme high sea level events. In much of the state, a deposit of quartz-rich sand covers the more complex sediments deposited during the Tertiary. As one traverses the state from north to south, the quartz content decreases and the carbonate content increases. Indeed, in extreme southern Florida, carbonate deposition dominated. The area to the north is a transitional zone and contains both types of deposits. It is evident that the region again served as a basin during the Pleistocene, as the Pleistocene deposits in this part of the state are thicker than anywhere else. In fact, it is believed that the thickness of Pleistocene deposits averages around 9 meters (DuBar 1991), but there are locations in south Florida where the Pleistocene deposits are 60–70 meters thick (Perkins 1977). Thus the region of south Florida continued as a marine lagoon well into the Pleistocene. There were, however, interruptions in this setting caused by sea-level changes. There is ample evidence that there were sea-level fluctuations that changed the region from a marine system to a freshwater system. During periods of terrestrial deposition and freshwater flow, karstification certainly occurred, and there is evidence of at least five sequences of exposure based on the formation of duricrusts in the sediments (Perkins 1977).

In the Holocene, a wide variety of deposition took place in south Florida. There is evidence of carbonate, siliciclastic, and peat deposition as well as evidence of incipient karst formation taking place. There are vugs and larger voids within some of the limestone, and solution sinkholes can be found. It is these sinkholes that are of most interest to sinkhole scientists studying south Florida.

Most attention given to Everglades restoration has focused on the reconstruction of the natural flow through the system. The natural flow brought water from the Lake Okeechobee region southwest to the Ten Thousand Islands area. Now, however, flow is largely diverted to the southeast for flood control and water supply. There certainly have been efforts to restore the natural drainage, but none of the efforts have fully accomplished the reconstruction efforts. Certainly the presence of sinkholes in the region must be taken into account when attempting to restore the region.

Before discussing specific sinkholes of the region, it is important to discuss the hydrogeology of the area. There are three main aquifers in the south Florida area: the surficial aquifer, the intermediate aquifer, and the Floridan Aquifer. The Floridan Aquifer, located beneath the Hawthorne Formation, is quite deep and is not utilized. It is brackish in this area. Above this aquifer is the intermediate aquifer. The water in this aquifer, found in southwest Florida, is housed within thick Miocene deposits of the Hawthorne Formation and supplies drinking water to the growing region of Fort Myers and Naples. The surficial aquifer, which is at the surface and in connection with canals, rivers, and lakes in southeastern Florida, is housed in a variety of different rock types in the region; the most significant source of water in this aquifer is found within the Biscayne Aquifer. It serves as the principal drinking-water source for Miami and other populous communities on the southeast coast. The aquifer is found in a variety of materials, including the Pamlico Sand, the Miami Limestone, the Fort Thompson Formation, the Anastasia Formation, and the Caloosahatchee Formation. In places, the Biscayne Aquifer is covered by a thin, confining unit made up of peat and marl (Miller 1997). However, the canals and drainage ditches that crisscross the aquifer have made the connection to the surface widespread, causing a high potential for drinking-water contamination for the metropolitan regions of south Florida.

Tree Islands of the Everglades

Many of the sinkholes of south Florida are expressed as tree islands (often found in circular or ovoid shapes in plan view) in the Everglades. There are a number of these features throughout south Florida. They can be true islands set within a water body, or they can be found as a group of trees within a marsh or upland. In central and northern Florida, tree islands can exist as cypress "domes," which are actually topographic lows inhabited by the water-loving cypress trees. Discreet tree islands surrounded by water are common within the Ten Thousand Islands area of extreme southwestern and southern Florida. Here, the tree islands form on topographic highs within the eastern fringes of Florida Bay. Tree islands in the marshy portion of the Everglades can form within paleo-sinkholes in the region. However, a variety of hypotheses have been proposed for the formation of tree islands that do not include a karst genesis. Indeed, there is some evidence that they do not form from karst processes at all (Mason and Van Der Valk 2002).

Before discussing the formation of the Everglades tree islands, it is worth reviewing their ecological significance. First of all, it is clear that these features were not always tree islands. They are likely succession features that formed as the islands filled with peat. In fact, analysis of the peat underlying the forested islands indicates that they were once open water and then sawgrass. Forests were relatively new features in the region. The peat stratigraphy shows open water species at the base of the peat column, sawgrass in the middle, and trees, particularly holly, at the top (Stone et al. 2002). It is believed that the mature tree island ecology has been in place for the last 500–1,200 years (Willard et al. 2002) and that they have long been an important ecological feature of the region (Orem et al. 2002). Although important for various nutrient cycles in the Everglades, the tree islands were also important for Native Americans. Many of the tree islands are also archaeological sites, providing evidence of human habitation and activity that dates to 11,000 BP and continues to the present day (Carr 2002). Indeed, present-day Seminoles and Miccosukees continue to use the tree islands for meeting a variety of needs (Carr 2002). In fact, as we will see, Carr (2002) proposes a human cause for the formation of some of the tree islands.

The tree islands contain unique plant and animal communities. It is believed that birds are partially responsible for the distribution of seeds to these communities (Gawlik, Gronemeyer, and Powell 2002). Red bay, sweetbay, red maple, bald cypress, pond cypress, strangler fig, and gumbo-limbo are common trees found on the islands, and coastal-plain willow, wax myrtle, dahoon holly, cocoplum, pond-apple, and buttonbush are all common shrubs found there (Lodge 1994). There is concern that alterations to the Everglades' hydrology over the last 100 years may have stressed these communities to their limits (Connor et al. 2002). Indeed, it is believed that all of the islands have been impacted in some way by a variety of disruptive factors including drainage, burning, logging, and the invasion of nonnative species (Armentano et al. 2002). It is likely the tree islands will continue to experience hydrologic change in the future (Heisler et al. 2002) and that the healthful role of fire in their ecology will remain significant (Wetzel 2002). Approximately 300 animal species are associated with the tree islands (Meshaka et al. 2002) including the green tree frog, the dusky pygmy rattlesnake, the anhinga, the roseate spoonbill, the great horned owl, white-tailed deer, and the Florida panther. Each species relies in some way on the islands for its survival (Brandt et al. 2002). Because the tree islands are so important to the life cycles of native animals, there is concern that their ecology will be significantly disrupted by the activity of nonnative animals, such as the feral pig. At the same time, the size and hydrology of the tree islands impacts animal ecology as well (Gaines et al. 2002).

Formation of Tree Islands in South Florida

There are two broad classes of tree islands in South Florida: floating and stationary. Floating tree islands form when a vegetative mass breaks away from a shore or other vegetative body. When this happens, the vegetative mat, typically a peat-based mat, serves as the host for the trees. This can happen in riparian settings, lakes, and wetlands. Many of the tree islands in the Everglades form when pieces of peat break away from the bottom and float to the surface. These types of islands are found most typically in the Arthur R. Marshall Loxahatchee National Wildlife Refuge (Van Der Valk and Sklar 2002).

The second type of tree island, the stationary or fixed tree island, forms under a variety of conditions. In a coastal environment, fixed tree islands can form on natural topographic highs such as an exposed reef bed or other elevated topographic feature. They can also form on anthropogenic deposits such as a prehistoric shell midden. These islands are subject to modification from waves and tides and can become significantly altered during major storm events. Tree islands in the interior of the peninsula in south Florida form differently. There are two main settings in which these features form: those associated with distinct topographic highs and those associated with distinct topographic lows. There are also settings where topography seems to have no relationship at all with the islands.

The elongated tree islands in the Everglades, somewhat teardrop shaped, are quite distinctive. The head of the teardrop lies to the north and is the highest point of the island. The tail of the island lies to the south at a lower elevation. This has led to speculation that the tails of these islands formed as depositional features caused by water flowing over the heads during flood events. While there may be some truth to this hypothesis, there is also evidence that the tails and the heads formed at the same time. Their linearity, therefore, might be caused by some type of topographic variation that existed prior to the formation of the islands.

Where the topography is high, the islands are found associated with distinct high points in the limestone bedrock surface or with mounds of peat. The tree islands associated with the mounds of peat probably started as floating tree islands that became stabilized as they became beached, as water levels fell, or as the mass of the peat increased over time to cause a thickening of the substrate. Regardless, these thick peat-mound fixed tree islands have no karst genesis. The tree islands associated with topographic highs in bedrock often have thin soils. The location of these bedrock highs is not well understood, although they could be subtle erosion remnants left from the downward solution of limestone. Only a few centimeters' change in bedrock elevation in the Everglades is significant and, thus, the tree islands that are bedrock based certainly are of geologic interest.

There are many tree islands in the Everglades associated with topographic lows. Among the many possible explanations for the formation of these depressions is that they could be solution sinkholes. The surface of the bedrock in the Everglades has a distinct patina that indicates a significant amount of chemical weathering. While much of the solution weathering would occur uniformly over an area, there are places where the solution could be enhanced, such as along joints or joint intersections in the bedrock, at places of recharge, in springs, and in mixing zones. Regardless of the reason, there are subtle depressions that form in the Everglades from karst processes. When these fill with sediment or organic matter, they provide a zone of well-drained material that serves as a perfect rooting place for shrubs and trees.

There is some evidence of the trees themselves enhancing the depression. The acids produced by trees and shrubs can enhance the solution of the limestone. In addition, the roots can work through joints, pores, and vugs in the limestone to further enhance the depression. Other processes can cause subtle depressions in the Everglades. Gator holes are one of several possible explanations for the formation of tree islands proposed by Stone et al. (2002).

Carr (2002) suggests that prehistoric humans may have been partially responsible for the formation of tree islands in the Everglades. There is evidence that some prehistoric peoples lived in homes built on poles. If a home were to fall, it could become a nucleus around which tree island vegetation could grow. Similarly, as suggested by Stone et al. (2002), animal nests, particularly those of rice rats, muskrats, sandhill cranes, and alligators, could serve as accumulation points for vegetation. These animal nests serve as topographic highs in the marshlands and thus can be places where, over years, detritus can collect to the point that shrubs or trees can take root.

A classification scheme for the tree islands of south Florida that divides the islands by geographic location and vegetative community was developed by Armentano et al. (2002) (see table 4.1). The tree island regions are divided into two broad classes: interior and coastal. The Interior Region is further subdivided into the Atlantic Coastal Ridge, the Big Cypress Region, Shark Slough and Adjacent Prairies, and Taylor Slough subregions. The Coastal Region is divided into the Lower Gulf Coast, Florida Bay, and Florida Keys subregions. Each subregion is subdivided again into several types, based mainly on vegetation community. This classification does not take into consideration the genesis of the tree islands. However, the "freshwater Indian midden" classification unit in the Big Cypress Region and the Lower Gulf Coast subregions indicates that human activity influenced the formation and development of the tree islands in these areas.

Given the detailed classification scheme presented by Armentano et al. (2002), it is intriguing to ponder whether or not there is some relationship between the vegetative communities and the genesis of the tree islands. In other words, does one community favor one type of setting over another? While we currently do not know the answer to this question, the classification scheme does provide a framework for future inquiry. This is particularly important as researchers develop models for understanding how the tree islands may react to future stresses in the Everglades as a result of altered hydroperiod (Wu et al. 2002).

What would greatly assist in the understanding of the threat to the tree island communities is an understanding of the substrate on which they form. Certainly the presence and thickness of peat deposits, depth to bedrock, hydraulic conductivity of layers beneath the subsurface, and other physical attributes influence the hydrology of the tree islands. Given that these factors are greatly impacted by the way an island formed, the genesis of the island is an important factor in the way the island responds to environmental stress. Perhaps, then, a survey of the substrate of the tree islands within each of the classification groups would assist in our better understanding their ecology. Furthermore, this survey could help researchers better understand the geologic and environmental history of the Everglades.

Regardless of how tree islands form, it is important to reiterate that they form from multiple processes and that there are many possible explanations for their formation. Yet many islands serve as clear evidence of sinkhole formation in south Florida. While not as topographically dramatic as the sinkholes that form to the north, these sinkholes serve an important function in the wetlands of south Florida. The solution sinkholes present in the Everglades are important to the regional hydrology. A better understanding of the sinkholes and the broader karst systems in the Everglades will augment the research that is being undertaken as part of the most recent Everglades restoration effort.

Also, as efforts move forward to restore the natural hydrology in the Everglades, it is important to take the karst system into consideration. Many unnatural modifications that have significantly impacted the natural flow of water have occurred in the Everglades. One of the largest of these is the result of drainage canals previously discussed. In addition, the lowering of land in agricultural regions of the Everglades is important. In these locations, the land has been lowered as a result of the oxidation of the peaty soil. Many feet of topography have been lost as a result of the drainage of these lands. Other modifications include the use of some of the Everglades for storm-water storage and the use of other areas as well fields. Regardless of the modification, it

Table 4.1. Tree island classification scheme

Region	Subregion	Type	Source
Interior	**Atlantic Coastal Ridge**	Tropical hardwood hammock	Olmsted et al. 1980; Horvitz et al. 1995
		Mahogany hammock	Olmsted et al. 1980
	Big Cypress region	Tropical hardwood hammock	Gunderson and Loope 1982
		Oak-sabal hammock	
		Freshwater Indian midden	Borel 1997
	Shark Slough and adjacent prairies	Tropical hardwood hammock	Oberbauer and Koptur 1995; Loope and Urban 1980; Armentano et al., unpublished raw data, 1995–97
		Bayhead swamp forest	
		Bayhead	
		Bayhead-hammock forest	Loope and Urban 1980
		Willowhead	
	Taylor Slough and Southeast Saline Everglades	Tropical hardwood hammock	Ross et al. 1996
		Bayhead	Jones et al., unpublished raw data, 1998
		Bayhead-hammock forest	
Coastal	**Lower Gulf Coast**	Tropical hardwood hammock	
		Strand hardwood hammock	
		Indian midden hammock	Borel 1997; Reimus 1997
	Florida Bay	Tropical hardwood hammock	Olmsted et al. 1981; Ross et al. 1996; Jones et al., unpublished raw data, 1998
		Buttonwood hammock	
		Thatch palm hammock	Jones et al., unpublished raw data, 1998
		Florida Bay Keys hammock	Jones and Armentano, unpublished raw data, 1996
	Florida Keys	Upper Keys hammock	Ross et al. 1992
		Lower Keys hammock	

Note: The table classifies tree island communities by region, subregion, type, and source (that is, the authors who studied or identified the tree island type). As one can see, there is a great diversity of tree island communities in south Florida.

Source: Modified from Armentano et al. 2002, 233.

is evident that natural recharge and discharge cycles have changed considerably (Harvey et al. 2002). Natural chemical cycles, such as that of mercury, have also been modified (Harvey et al. 2002). We do not know how all these changes have influenced karst processes or how the subtle karst features of the Everglades have affected natural flow, discharge and recharge cycles, and geochemical cycles. A great deal of research needs to be conducted on these topics in order to better elucidate the karst of this important region.

5

Notable Sinkholes

Florida is home to many unusual or significant sinkhole, but several stand out. The Winter Park Sinkhole, which originally extended 300 meters in diameter, is certainly the largest and most dramatic of the recent sinkholes. It exemplifies the power of karst processes in our landscape and points to the tremendous void space in the bedrock under our feet. Another sinkhole, Devil's Millhopper in Alachua County, is a unique feature that distinctly charts our geologic past. Paynes Prairie, also in Alachua County, is a large coalescing sinkhole system that was very important in the history of our state. There are also several sinkholes that formed in recent years as a result of longstanding drought. These point to the fact that sinkholes are very much influenced by our weather, climate, and hydrology. Each of the sinkholes or sinkhole groups described below helps us understand the geological and human history of our state in distinctive ways.

Winter Park

The most dramatic sinkhole event in modern Florida happened in Winter Park, near Orlando, in 1981. The sinkhole formed at the corner of Fairbanks Avenue and Denning Drive. It is quite easy to visit the site. No matter the point of departure, the best way to get to the sinkhole is via I-4. North of Orlando, take exit 87 onto 426, which is Fairbanks Avenue. Travel east on Fairbanks several blocks to Denning Drive, and then turn left on Denning Drive. The sinkhole will be the first lake you see on your left. Today, the site of the Winter Park Sinkhole is bounded by businesses on the Fairbanks Avenue side of the lake and by parkland and sidewalks on Denning Drive and opposite the Fairbanks Avenue side.

The collapse occurred rapidly, over the course of a single day, in 1981. By the end of the day, a swimming pool, a car dealership, and several buildings had been destroyed. The event reminded Floridians of the instability of the ground surface and prompted new laws that I will discuss in chapter 7. National and local news covered the story of the collapse for days. Even today, the photograph of the sinkhole shortly after it formed is used in introductory geology and physical geography textbooks. A simple Internet search for "Winter Park Sinkhole" produces numerous websites, illustrated with dramatic photographs, which discuss its formation and aftermath. Some websites of companies that focus on sinkhole law or property remediation also utilize photographs of the sinkhole.

Immediately after collapse, the sinkhole measured 106 meters in diameter and was 30 meters deep. The engineers who examined the feature after it formed (Jammal and Associates 1982), reported that although the formation was spectacular, it had been forming for years as overlying sands raveled into voids in the subsurface. In fact, Jammal believes that its formation was accentuated over the last 50 years due to the steady drawdown of the piezometric surface of the Floridan Aquifer (Jammal 1984). He notes that the piezometric surface was at +20 meters in the 1930s and that in the 1980s it was at +14 meters, a drop attributable to below average rainfall and extensive well pumping (Jammal 1984, 364).

After the sinkhole formed, it took several weeks for the natural formation to stabilize. Initially, a great deal of debris was sucked into the subsurface, including a house. Finally, the sinkhole plugged and the depression filled with water. However, in the weeks following the formation, the sinkhole drained suddenly twice, indicating that the plug had been breached and that water could flow freely into the subsurface Floridan Aquifer. After each breach, the sinkhole quickly plugged again and its hydrology stabilized (Jammal and Associates 1982).

Once the land stabilized, the City of Winter Park redeveloped the site by stabilizing the edges near the businesses on Fairbanks Avenue and along the road on Denning Drive. The area that was once part of a pool is now a ball field. The site needed extensive grading and demolition expertise in order to restore the urbanized landscape.

The formation of the large sinkhole in Winter Park is a rare event but one that should not have been unexpected. There are numerous small- and mid-sized lakes dotting the area, and most of them are round and clearly sinkhole lakes. Some of the lakes have hourglass or more complex forms, indicating that they formed from complex multiple sinkhole occurrences. As Jammal noted in his study of the Winter Park Sinkhole, a decreased piezometric surface can induce sinkhole formation on the scale of the Winter Park Sinkhole. As Florida goes through boom-and-bust periods of natural rainfall and experiences growing demands on its aquifers from a growing population, it is likely we will see more such sinkholes form in the future.

Paynes Prairie

Paynes Prairie is a broad wetland south of Gainesville in Alachua County. Its unique history is intimately tied to major historical events in the whole of Florida. The history and use of the prairie was dependent upon the active karst and hydrological activities that took place there.

The easiest way to access Paynes Prairie is to exit I-75 at Micanopy. South of the site, the town of Micanopy also has a rich history. Many of its original nineteenth-century buildings are still standing, and the downtown area is known for its antique stores. The charming setting of the town within easy driving or biking distance of the prairie makes the area an excellent day-trip destination, especially for those interested in natural and regional history as well as some shopping.

In 1971, the state designated 22,000 acres of the land around Paynes Prairie as the Paynes Prairie Preserve State Park. There are miles of biking and hiking trails through the park, and there is a visitors' center on the southern edge of the prairie within the park. The visitors' center, which has restrooms and a small gift shop, contains several interpretive displays on the geologic and geomorphic history, ecology, and human history of the park from prehistoric times to the present. The center also serves as a scenic overlook into the beautiful prairie landscape. Within an easy walk from the center is a viewing tower. Rising several stories, the tower provides a panoramic view of much of the prairie (figure 5.1). In addition, the visitors' center shows a half-hour informative video on the natural history of the prairie that is suitable for all ages.

Paynes Prairie has a rich history that dates from the earliest Native Americans in the region through the Spanish exploration and into the modern era. Each culture modified the prairie in some way, and it is worthwhile to consider the broad cultural history of the prairie before reviewing its geology. Anderson (2004) provides a detailed history of the prairie, and the reader is referred to his work for a more comprehensive view.

Evidence of early Native American presence in Florida, including the land in and around Paynes Prairie, abounds. While the earliest human inhabitants

Figure 5.1. Paynes Prairie in Alachua County, home to a large, coalescing sinkhole system.

of the state date to about 12,000 years ago, the first significant known landscape alteration around Paynes Prairie occurred approximately 2,200 years ago. There is evidence of human habitation near the prairie prior to this time, but it is largely in the form of lithic and pottery remains. Yet, a little over two millennia ago, a culturally identifiable group began to live around Paynes Prairie. It is thought that at the time of settlement, the prairie was a lake. The people who settled around it, called the Cades Pond people, lived mainly on game and freshwater fish and built a series of mounds around Paynes Prairie, particularly on its eastern side. The mounds appear to have served two purposes: one largely ceremonial and the other funerary (Anderson 2004). Though many of the mounds have been destroyed, a few remain intact.

A new culture developed around the year 800 that brought widespread agriculture to the region. Instead of relying mainly on game and fish, this people planted a variety of crops. They were similar in many respects to the Native American groups found in southern Georgia at the same time. Cultural diffusion or invasion might have driven the transition of the Cades Pond culture to a new group that anthropologists call the Alachua Tradition cultures. Interestingly, mound building in general ceased during this period.

The people of the Alachua Tradition evolved into the culture that was present at Paynes Prairie at the time of European contact: the Timucua. The Timucua inhabited a wide swath of Florida north of Tampa Bay and east of

Tallahassee including Alachua County. While culturally similar, tribal groups within the Timucua were nonetheless distinct. The Paynes Prairie Timucua were known as the Potano. In 1539, Hernando De Soto began his long trek across the southeastern United States, beginning in Charlotte Harbor. On August 12, he came into contact with the Potano near Paynes Prairie and spent the night in their main village. Beyond this record, there is very little known of the group that lived around Paynes Prairie at the time of De Soto's visit. Indeed, it is currently not known where the main Potano village was located, although it was probably on the southern shore of the prairie, perhaps a few miles from present-day Micanopy. There was much discord between the earliest European settlers and the Potano in the mid-sixteenth century. The French, who had settled south of St. Augustine, tried to gain regional control by allying themselves with the Outinas, an indigenous group not on friendly terms with the Potano. Their battle against the Potano met with only limited success. The French retreated and were soon massacred by the Spanish, who also later used the Outinas to attack the Potano—with disastrous consequences for the Spanish.

It was not until the 1600s that formal arrangements with the Potano and other Timucua allowed for the development of a series of missions across north Florida to secure Spanish cultural dominance. The Mission San Francisco de Potano was established north of Gainesville with a mission outpost on the south shore of Paynes Prairie. Close contact with the Spanish devastated the Potano and the other Timucuan peoples. Many Native Americans died from diseases brought to the area by Europeans, and Indian cultural traditions were lost as Native settlements drifted toward the missions. In all, what was a cultural death knell for the Potano enabled the Spanish to begin dividing the Potano lands for Spanish ownership.

The first person to develop the lands around Paynes Prairie was Francisco Menéndez Marqués. A descendent of Pedro Menéndez de Avilés, the founder of St. Augustine, Francisco began a cattle ranch in the vicinity that remained active from the 1640s through 1656. At that time, the remaining Potano rebelled against the Spanish. The Potano had been used as slave laborers, forced into many of the tasks associated with cattle ranching. As part of the revolt, they killed the cattle. In 1670, Francisco Menéndez Marqués returned to Paynes Prairie and decided to live near its northern edge at Alachua Sink. There, he rebuilt his ranch, which he called La Chua or "the jug." This name refers to the sinkhole, or jug, near where Francisco sited his ranch.

The lands around Alachua Sink were perfect for bovine herding, and La

Chua prospered. In fact, the ranch was so successful that pirates invading from the Gulf Coast twice attacked it. Eventually, the ranch was not able to compete in the international markets, and, as St. Augustine was the hub of the cattle trade in Florida, the Menéndez Marqués family largely abandoned the Alachua site. Later, the area again came under attack by the English as they asserted their dominance over the southeastern United States in the early 1700s. At this time, the English had a take-no-prisoners approach to the Native peoples and slaughtered hundreds of them. In addition, they burned Spanish missions and attempted to eliminate Spanish influence in the Southeast.

The English were allied with the Creek Indians who lived north of the Timucua. The Creeks were encouraged by the English to attack the north Florida Timucua—many of whom lived around the missions—with the goal of removing the Spanish altogether from the region, particularly St. Augustine. After much loss of life in the early 1700s, the Spanish left the area around Paynes Prairie in 1706, leaving the region open to other settlement. The abandoned lands were then settled in the mid-eighteenth century by a Creek tribe called the Oconee, who probably occupied the original Potano village site. At this point, the karst nature of the landscape plays a role in the history of the Oconee settlement. They originally farmed, hunted, and raised cattle on the land. However, shortly after their arrival, the sinkhole plugged and the region around Paynes Prairie flooded. This expanded the resource base of the Oconee, but several years after the lake formed, the sinkhole plug drained and the lake disappeared. The stink from the dead fish caused the Oconees to move their village to the present site of Micanopy.

At the close of the French and Indian Wars in 1763, the Spanish ceded control of Florida to the English. The Oconees, under the leadership of the tribal leader, Cowcatcher, hated the Spanish and forged an alliance with the English. It was at this time that the Oconees became known as the Seminoles. In addition, La Chua became known as Alachua Prairie.

In June 1774, the famous natural historian William Bartram visited Alachua Prairie and the Seminole village of Cuscowilla near present-day Micanopy, where he was well received by the Seminoles and Cowcatcher.

> The extensive Alachua Savanna is a level, green plain, above fifteen miles over, fifty miles in circumference, and scarcely a tree or bush of any kind to be seen upon it. It is encircled with high, sloping hills, covered with waving forests and fragrant Orange groves, rising from an exuberantly

fertile soil. The towering Magnolia grandiflora and transcendent palm stand conspicuous amongst them. At the same time are seen innumerable droves of cattle; the lordly bull, the lowing cow and sleek capricious heifer. The hills and groves re-echo their cheerful, social voices. Herds of sprightly deer, squadrons of the beautiful, fleet Seminole horse, flocks of turkeys, civilized communities of the sonorous, watchful crane, mixed together, appearing happy and contented in their enjoyment of peace, 'til disturbed and affrighted by the warrior of man. (Bartram 1792, 185–86)

Although Bartram's language embodies the effusive romantic literary style of his era, it is nonetheless clear that he was genuinely enthralled by the uniqueness of the landscape and the diversity of its plant and animal life. This is a reaction not unfamiliar to the present-day visitor to Paynes Prairie.

Eventually, the Spanish regained control of Florida after Queen Anne's War to the dismay of the Seminoles and their chief, who died shortly after this event. The new chief was Cowcatcher's son Payne, who took the title King Payne. Throughout King Payne's rule of the Seminoles around Paynes Prairie, a number of Creeks migrated into north Florida from Georgia and Alabama, where they were driven out or treated as slaves. In addition, a number of runaway black slaves made their way to Florida and lived with the Seminoles and the Spanish, establishing communities at Fort Mose near St. Augustine and at Paynes Prairie.

The presence of runaway slaves and Seminoles on the Georgia and Alabama borders with Florida made the state's slaveholders uncomfortable. In addition, the landowners of the Deep South saw the Indian lands as potentially highly profitable and sought to expel the Seminoles from their lands and gain control of Florida's Spanish frontier. In 1812, Colonel Daniel Newnan invaded the Seminole lands and did battle with King Payne. A strong defense by the Seminoles caused Newnan to advance in force. Wisely fearing retribution from the Americans, the Seminoles left the Paynes Prairie area and divided into two bands. One band went west with Payne's brother Bowlegs, and one band went south with Payne, who died shortly after from a wound he received while fighting Newnan. As a footnote to the story of the Seminoles from Alachua Prairie, Bowlegs ended up as the leader of the tribe and fought with the British at the Battle of New Orleans against Andrew Jackson. Jackson later fought the Seminoles in the First Seminole War in 1818. Although the Seminoles remained in Florida for the time being, whites began to settle in the vacated areas. Even though the region went through more turmoil during the Second Seminole War (1835–42) and the Civil War (1861–65), settlers began to clear land and develop the region throughout the nineteenth century. Orange groves and cotton fields were common in the uplands surrounding Paynes Prairie, and the savanna was used for grazing cattle. As agriculture expanded, the region's economy developed.

In 1867, things began to change for the worse. Alachua Sink became plugged again, and the savanna flooded. The exact cause of the plug is unknown, though it certainly could have been the result of some natural event. At the time, however, it was believed to have been caused by human activity. Even then, the sink was a tourist destination. People would come from neighboring communities to see how water drained at that location directly into the earth. Some visitors threw detritus such as logs and brush into the sink to watch it drain to the bottom and vanish. Unfortunately, this seemingly playful activity may have contributed to the plugging of the sink and to the widespread flooding that followed.

The settlers were aware that the prairie flooded on occasion, yet the water always went down again as it drained through the surface or overland into Alachua Sink. But in 1867, the water didn't drain. In fact, it got deeper and deeper and created a lake called Alachua Lake. The formation of the lake led many to think about how to modify the prairie for their own purposes. This set off a century-long series of modifications that changed the ecology, hydrology, and overall quality of the prairie forever.

These changes centered around two main and conflicting themes: (1) that the prairie remain permanently flooded to allow for freshwater fishing, boating, and water front property development and (2) that the prairie be drained to allow for agricultural activities, particularly cattle grazing. Numerous plans were hatched over the years to realize one or the other of these visions. Interestingly, such debate remains central to Florida land use, where real estate and tourism interests frequently vie with agricultural interests.

Although many people wanted the lake drained after it formed in 1867, no modifications to the prairie were undertaken. In fact, the region adapted to its new lake. A small steamship even transported agricultural products from one side of the lake to the other. The adaptation to open water was short-lived, however. The lake drained over the course of ten days in 1892. Vegetation soon covered the former lake bottom and the wet prairie returned, as did the cattle. But the land was still wet, and many people thought it needed further draining to make it more profitable. Thus a complex series of canals and dikes were built in and around Paynes Prairie in order to fully drain the system and allow widespread agricultural development. A series of roadways were added and the natural flow of water from the broad polje into Alachua sink ended. Although agriculture did expand in the prairie after it was drained, it never attained the scope imagined by the individuals who wanted to drain it. Thus, as concern over natural lands grew in the twentieth century, there was interest in preserving the lands and restoring the prairie.

The state of Florida purchased most of the prairie lands in 1970. Since that time, major restoration efforts have been undertaken, although much of the hydrology is still altered due to the dikes and canals that remain in place to protect private landholdings within the prairie.

The history of the prairie has been greatly influenced by the Alachua Sink. Over the years, the sinkhole plugged and unplugged several times. Each time this happened, land use, and thus the history of the region, was affected. The sinkhole was regarded as a blessing by some and as a curse by others. When the sinkhole plugged, the disruption of agricultural activities provided opportunities for those with competing interests. When it drained, the interests of agriculture but not of recreation were served. Perhaps now that it is largely in state control, the sinkhole can plug or unplug without causing too much turmoil.

Devil's Millhopper

Devil's Millhopper is a striking sinkhole located northwest of the city of Gainesville in Alachua County (figure 5.2). The reason the sinkhole is so fascinating is that it has the greatest local relief in the region. In fact, the land around the feature is distinctly flat. The park around Devil's Millhopper, easily found off I-75, is currently a state geologic site. The site has a small visitors' interpretive center and restrooms. Past the visitors' center is a hiking trail, one leg of which traverses a circular path through the upland surrounding Devil's Millhopper and another that extends into the depression. The latter trail descends a series of steps to the bottom of the sinkhole. During the wet season, small waterfalls cascade down the fern-covered sides of the sinkhole. (One needs to be in reasonable shape to make the climb back up the stairs out of the sinkhole.)

The site where the sinkhole is located is on the northeastern fringes of a karst-dissected plain and a flat upland. The sinkhole itself is in an area that is transitional between the two regions. There are very few surface streams

Figure 5.2. Devil's Millhopper, a striking sinkhole northwest of Gainesville in Alachua County.

in the area. Blues Creek and Turkey Creek, the only streams found in the Gainesville West Quadrangle where the Devil's Millhopper is located, flow from spring heads north of the Devil's Millhopper site. Both streams flow in a northwest direction in karst valleys to the San Felasco Hammock State Preserve, where Sanchez Prairie is located. Sanchez Prairie is diagenetically similar to Paynes Prairie. Numerous sinkhole lakes and wetlands are found in the vicinity of Devil's Millhopper, so the presence of a particularly dramatic s in this area isn't unexpected. However, the topographic expression of the feature is notable. Within less than one-sixteenth of a mile, the elevation drops from 175 feet to 55 feet (West 1973). While this 120-foot elevation change does not seem especially notable on a global scale, it is a remarkable topographic feature in the very flat state of Florida.

The karst-incised region to the southwest is different from the area around Devil's Millhopper due to the thickness of the overlying sediments. The region to the southwest has a topography similar to that of places like the Ocala Ridge and the Brooksville Ridge. What these places have in common is that each has a very thin covering of sediments atop the underlying limestone. In the area southwest of Devil's Millhopper, the Ocala Limestone is extensively carved with numerous sinkholes, uvalas, caves, and karst valleys. The relatively flat plain to the northeast of this area, where Devil's Millhopper is located, has a thick sequence of sediment on top of the Ocala Limestone. In fact, the sides of Devil's Millhopper do not expose any limestone. Instead, sediments of the Hawthorne Formation are exposed to the base of the sinkhole, where the Ocala Limestone is exposed.

The relative closeness of the thick body of sediments adjacent to the carved-up karst to the southwest indicates that the site is near the horizontal contact of the Hawthorne Formation to the northeast and of the Ocala Limestone to the southwest. This contact signals that the area was once the margin of a basin in which the Neogene Hawthorne Formation sediments were deposited. Interestingly, the topographic highs are similar on either side of the contact. As mentioned above, what is striking about Devil's Millhopper is its extreme local relief. Sinkholes that form where there is a thick cover of sediments are rare, but where they occur, they are deep and steep-sided. Conditions in the vicinity of Devil's Millhopper are perfect for the formation of deep, steep-sided sinkholes that contrast sharply with the broad, deep sinkholes in the thinly covered Ocala Limestone to the southwest.

Pinellas County Sinkhole Swarms

Pinellas County, the most densely populated county in Florida, has experienced several sinkhole swarm formations that have been the subject of much concern to the public and the media. Pinellas County, a peninsular coastal county located in west-central Florida (figure 5.3), grew up around the original settlements of St. Petersburg and Tarpon Springs and expanded to include Clearwater, Largo, and several other communities. There are very few undeveloped land parcels in the county, and the entire area is urbanized. Geologically, the county is underlain by surficial Quaternary undifferentiated

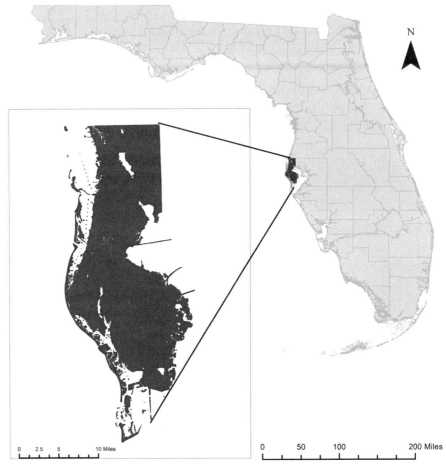

Figure 5.3. Magnified portion of northern Pinellas County.

sands and clays, which are underlain by the more complex Miocene Hawthorne Group sediments that include the clayey Peace River Formation and the Arcadia Formation. Beneath the Hawthorne sediments are the Oligocene Suwannee Limestone and the Eocene Ocala Limestone. The landscape of the county is quite striking. It has a barrier island complex along the Gulf of Mexico coast and a mangrove tidal swamp on the shores of Tampa Bay. In the interior, a series of ancient beach terraces extend to elevations of over 100 feet above sea level. Sinkhole lakes are found throughout the county.

Reports of sinkhole damage to homes are common in the county. As a rule, the damage is subtle: cracked foundations or cracked plaster. The county's geology makes it hard to determine whether a sinkhole is responsible for particular damage because there are three main natural causes of ground instability in Pinellas County: sinkholes, shrink-swell clays, and the oxidation of peat. In particular parts of the county, shrink-swell clays occur as lenses in the Hawthorne Formation. Unfortunately, shrink-swell clays are known to cause structural damage in areas also prone to sinkhole activity, so linking specific damage to one or the other cause is often hard to do. In addition, there are many at depth peat deposits in Pinellas County. If the water table drops below the level where the peat is located, the organic matter can oxidize and the land can subside. This triple threat posed by sinkholes, peat, and clay makes the area particularly fascinating for those interested in urban geology.

One of the problems associated with fully understanding the sinkhole phenomenon in Pinellas County is that for a long time no one had a good understanding of sinkhole formation and distribution prior to land development. In fact, urbanization of the area destroyed or covered numerous sinkholes. In order to assess the extent of sinkhole formation prior to urbanization in the county, Wilson (2004) examined aerial photos from the 1920s. Pinellas County is one of the few areas in the nation where such early aerial photos were taken. Using them, Wilson mapped circular depressions throughout the county prior to widespread development and compared the results to maps she created by mapping circular depressions on recent aerial photos. She found that 87.13 percent of topographic depressions were lost to development over this seven-decade period. What this means is that hundreds of depressions were filled or in some way destroyed as the county grew. Indeed, many structures were built on existing sinkholes. While most of these sinkholes are dormant, some may have been active at the time of development and may remain active. This phenomenon is certainly not unique to Pinellas County. There are probably dozens and dozens of neighborhoods in Florida that were built on filled sinkholes. The 2,703 sinkholes found in 1926 made up approximately 44 square kilometers of the county and 6 percent of the overall land area. By 1990, only approximately 6 square kilometers of land were classified as sinkholes, making up 0.8 percent of the total land area.

Most of the sinkholes located in the 1926 aerial photos lie in the northern half of the county. Sinkholes also showed up in the southern half of the county, particularly west and northwest of downtown St. Petersburg, but they were not as numerous as those in the northern half (figure 5.4). Interestingly, the sinkholes in the northeastern portion of the county are the most complex and show clear signs of multiple formation processes. Indeed, it appears as if they may be in some way related to dune or other coastal features. The sinkholes Wilson mapped most recently are mainly located in the northern portion of the county, particularly the northeastern section. In 2000, very few sinkholes existed in the southern portion of Pinellas County. Wilson indicated several exclusion zones where she was unable to conduct analyses using either the 1926 or the 2000 photographs because some of the areas had already been urbanized by 1926, particularly Tarpon Springs and St. Petersburg. In addition, she found that portions of some of the aerial photos were of low quality and not useful for her analysis.

Wilson also found that a large number of Pinellas County sinkholes have been turned into water-retention ponds. This is problematic because such ponds are designed to hold water for long periods of time to prevent polluted storm water from entering the aquifer. To slow drainage, such ponds are typically lined with some type of low-permeability material, usually clay. If a sinkhole is present in the pond, the raveling zone of the sinkhole can serve as a pipe that connects the subsurface water body with the Floridan Aquifer, bypassing the filtering effects of the surface sediments and the slowly permeable Hawthorne Formation. Wilson mapped 1,646 storm-water retention ponds on the recent aerial photos. She was able to identify these features by their artificial characteristics—constructed sides, unnatural, irregular forms—and by the fact that they were not present in the 1926 aerial photos. The actual number of water-retention ponds in the county no doubt exceeds what Wilson was able to document, since very small features fell outside the visible scale of the aerial photos she utilized. Of the 1,646 water-retention ponds she mapped, 499 of them, or nearly one-third, were sinkholes in 1926.

If we examined sinkhole conversions in other urbanized portions of Florida, we would probably find similar results. Many sinkholes throughout the state have been converted to water-retention ponds and many others have been filled and developed for urban land use. Risk is associated with both of these activities. If a sinkhole remains active after being filled, building foundations on top of it can crack and the property can be significantly damaged. Also, sinkholes utilized as water-retention ponds can allow polluted water to drain directly to the aquifer. Current regulations prohibit the use of sinkholes as water-retention ponds, but there are hundreds of sinkholes in Pinellas County that still pose such a risk, and it is likely that thousands of water-retention ponds throughout Florida are a threat to the water quality of the Floridan Aquifer.

In 1991, Beck and Sayed examined the geology of Pinellas County and the overall root causes of the sinkhole problems in the area. In addition, they investigated forty-two sinkhole occurrences in Pinellas County that were part

Figure 5.4. Sinkhole map of Pinellas County, 1926. Adapted from Wilson (2004).

of the Florida Sinkhole Research Institute's database on Florida sinkholes. However, they did not investigate the reported incidents of land instability that occurred in 1990 and 1991 in the Patricia Estates area of Dunedin. Here, dozens of insurance claims were made, and in many instances it was determined that shrink-swell clays were responsible for the damage.

Of the forty-two sinkholes investigated in their study, Beck and Sayed found that most of them, 34 to be exact, were of the cover-collapse variety while the remainder were cover-subsidence sinkholes. Cover-collapse sinkholes form rapidly as the cover over a limestone void gives way. Although they form rapidly, the processes leading to formation may take many years as sediment ravels into the voids without altering the surface expression of the landscape. Cover-subsidence sinkholes, in contrast, form slowly as sediments ravel into the subsurface. Like an hourglass, the surface of the landscape slowly indents as the sediment filters into voids in the limestone. The topographic expression of each of these sinkhole types is different, at least initially. When it first forms, the cover-collapse sinkhole has steep sides. With time, the sinkhole's edges slump into the depression to form a bowl-shaped depression similar to the expression of a cover-subsidence sinkhole. Thus, over time, the two sinkhole types cannot be distinguished based on topographic expression alone.

It might be that the large number of cover-collapse sinkholes documented in the area is a result of reporting bias. Cover-subsidence sinkholes form slowly and are not particularly dramatic. They would likely not be reported unless they were to cause significant damage. However, a sudden collapse of the surface is always a notable event, so cover-collapse sinkholes probably have been reported more frequently.

Frank and Beck (1991) examined similar sinkhole-related problems in Dunedin, where 171 homes were reportedly damaged by some kind of subsurface movement. There was public debate about the actual cause of the damage, given the three possible sources of ground instability: sinkholes, peat oxidation, and shrink-swell clays. In this case, the majority of the damage in the Patricia Estates area was found to be the result of sinkhole formation. While Frank and Beck based this conclusion on many pieces of evidence, two are signally important: (1) ground-penetrating radar indicated the presence of sinkhole activity in the subsurface and (2) round surface depressions were found in yards where homes had suffered some type of damage. The authors also report that structural cracks did not open and close as would be expected with shrink-swell clays, and borehole tests confirmed the presence of raveling zones. Interestingly, Frank and Beck concluded that a municipal well, City Well 5, a source of concern due to high groundwater withdrawals, was not responsible for subsurface instability. They found instead that the well had minimal impact on the regional water table and that there was no evidence it was causally linked to property damage in the Patricia Estates area. The authors further point out that many homes adjacent to the well were not damaged as a result of pumping and that it would be illogical to assume that homes distant from the these properties would be adversely impacted by pumping.

Although scattered across the county, cover-collapse sinkholes in Pinellas County are generally concentrated in the northern portion between Tarpon Springs and Lake Tarpon. The region where there is the greatest concentration is also the portion of the county where the sediments above bedrock are thinnest. Beck and Sayed (1991) delineate this area as having the greatest risk of cover-collapse sinkhole formation. In fact, they calculate that there were 15 sinkholes in 19 square miles over a 16-year period. At this rate, it can be anticipated that a sinkhole will form in any square mile of the county every 20 years. They further calculate the risk of damage to property in the area, estimating that there is a 1 in 5,555 chance of a home's being damaged once every 50 years. While this may not seem a particularly high risk, it should be noted that, over the last several decades, cover-collapse sinkholes in the region have damaged several homes. It should also be noted that these sinkholes are not distributed evenly over time. In fact, their occurrence is clustered over particular years or groups of years. In addition, such sinkholes most typically form in April, with two-thirds occurring between January and April. The fact that they occur during particular seasons and over particular years indicates that there is probably some climatological variable that influences the formation of sinkholes. Even so, sinkholes can occur any time, and climate is not the sole factor driving their formation.

Most of the cover-collapse sinkholes that formed over the course of the Frank and Beck study period are quite small, typically less than 6 meters in their longest axis. Only a few exceed this size. What this means is that Pinellas County sinkholes are usually troublesome events of small magnitude and very different from the more dramatic sinkholes found at Winter Park. Instead, when the sinkholes here form, they may or may not damage structures, depending on whether they intersect with developed portions of the landscape. Indeed, most of the damage is on a small scale: cracked foundations, damaged roadways, or some other repairable problem. On occasion, however, the sinkholes form in locations where property is destroyed. Of course, they do pose a risk to those living in the area if they are in the wrong place at the wrong time.

As noted above, there is some degree of regional expression of cover-collapse sinkholes in Pinellas County. This is not the case for the cover-subsidence sinkholes mapped by Beck and Sayed (1991). Only eight sinkholes of this variety are recorded, so it is not entirely unexpected that they show no distinct clustering pattern. In fact, these sinkholes were typically small features less than 6 meters in diameter. There is some debate as to what the cover-subsidence features in Pinellas County really are. Are they truly sinkholes, or are they some other settlement feature caused by shrink-swell clays or peat? The answer to these questions is important because property damage caused by sinkhole activity is covered by insurance while damage resulting from shrink-swell clays or oxidizing peat is not.

Sinkhole formation in Pinellas County is exacerbated by water drawdown. As noted elsewhere in this book, the depression of the regional groundwater table can trigger sinkhole formation. There is ample evidence of this occurrence in the agricultural regions of Hillsborough and Polk Counties during freeze events, when farmers spray thousands of gallons of water onto crops to protect them from frost damage. The northern portion of Pinellas County, near Dunedin, was pumped heavily in the second half of the twentieth century to meet the demands of the region's growing population. The pumping coincided with several sinkhole swarms, and there was considerable debate as to whether or not the pumping was responsible for the swarms and associated property damage. Evidence points to the likelihood that multiple factors caused the Pinellas County sinkhole swarms: deposits of shrink-swell clays in the vicinity of the well fields and voids in the limestone beneath the surface. While regional drawdown of the water table has certainly had some impact on the county's karst processes and shrink-swell clays, it is impossible to define a clear causal relationship. Multiple factors influence the landscape in the region.

Stewart et al. (1995) used a Geographic Information System (GIS) to determine whether there is any relationship between non-sinkhole features and foundation failures in the county in order to evaluate the range of causes of property damage linked to unstable land. The researchers examined the following spatial variables as GIS coverages: 1926 wetlands, National Wetlands Inventory wetlands, and soils. They attempted to determine if there was a relationship between those variables and 809 foundation failures (they obtained the foundation failure data from the Pinellas County Tax Assessor's Office). Interestingly, a significant number of foundation failures were located near the Patricia Avenue area, the portion of Dunedin already noted as having a high risk of foundation damage. In addition, data on sinkhole formation was included from the sinkhole database that was at the time kept by the Southwest Florida Water Management District. The 1926 wetlands were digitized from 1926 aerial photos, and the National Wetlands Inventory map was obtained electronically from Pinellas County. Two main soil types pose a risk to foundations, and both were utilized in building the model: fellowship soils are a specific local soil type that contain shrink-swell clays, and histosols are a broad soil unit that contains mainly organic matter that can oxidize when the soils are drained.

In order to assess if wetlands or soils have a relationship with foundation failures, the map coverages described above were placed within a GIS and compared with the distribution of foundation failures. This was done on a grid system to see if there was any relationship between the foundation failure and any of the other variables. The authors used 500-, 250-, and 125-square-meter grid sizes. Unfortunately, they were only able to complete the analysis on the northern half of the county for several reasons: the foundation failures are mainly in the northern half of the county; the 1926 aerial photos show that development had already occurred in the southern half of the county; the quality of the 1926 aerial photos used for wetland delineation was higher for the northern half of the county; most of the histosols and all of the fellowship soils lie in the northern half of the county; and the thickness of the overburden on top of the limestone is thinnest in the northern half of the county, making the formation of sinkholes there more likely. In other words, multiple factors are present that could induce foundation failure in the northern half of Pinellas County, whereas the risk factors in the southern half are significantly fewer. This conclusion is supported by the fact that fewer foundation failures are reported for the southern part of the county.

It was found that only the fellowship soils, which are shrink-swell soils with montmorillinitic clays, have a strong positive correlation with foundation failure. Histosols do not have any correlation with foundation failure. This is not totally unexpected because histosols are surface wetland features, and it is unlikely that anyone would have opted to build a home on the site of existing wetland organic matter. Nevertheless, unmapped areas of buried peat in Pinellas County still pose a threat to foundations there, because peat can oxidize and decrease in volume if the water table drops. Also, Stewart et al. did not find a link between the wetlands mapped by the National Wetlands Inventory and foundation failures. This is not unexpected because the wetlands existed in 1990, and it is unlikely that anyone has built structures on top of them since the map was created. However, a weak link connects wetlands mapped in 1926 to later foundation failures. Though not strong, this relationship nonetheless indicates that building on former wetlands, which are often sinkholes, poses some risk of foundation failures.

When the Stewart researchers examined the Patricia Avenue area of Dunedin closely, they found that sinkholes, not the presence of any particular soil type or wetland, were largely responsible for foundation failures. This is interesting in that the area is also the site of a municipal water well that caused a notable cone of depression. Some believe that the well initiated the sinkhole cluster in the Patricia area. However, another cone of depression associated with another well was also situated in the Patricia area, and buildings on that site did not experience any foundation failures. Clearly there are multiple factors that influence sinkhole formation and foundation failures.

A complete explanation of the nature of sinkhole formation and foundation failures in Pinellas County is not available at this time. What we do know is that sinkholes form with some regularity in Pinellas County, particularly in the northern section, where the thickness of the cover above bedrock is thinnest. Although there are certainly some human activities that can induce sinkholes in the region, such as extensive pumping, it is impossible to lay blame on a particular activity when so many sinkholes form as a result of natural processes. It is also clear that many sinkholes in the area were buried over the course of the county's urban development. How many of these are active sinkholes we will never know. Thus, it is impossible to determine whether the burial of sinkholes induces further sinkhole development or whether the buried sinkholes are responsible for property damage. Finally, there are other processes that mimic sinkhole damage to property. Shrink-swell clays and the oxidation of organic matter can cause specific damage to homes. Due to all of these factors, there is a great deal of professional disagreement about the cause of foundation failures in the region. Because only sinkholes are covered in homeowners' insurance policies in Florida, many geotechnical firms that specialize in determining the causes of foundation damage operate in the Pinellas County and the surrounding area.

Lake Okeechobee—Not a Sinkhole Lake

Lake Okeechobee is not a sinkhole lake although many people mistakenly believe it's a result of karst processes. I mention this fact here just to clear up any confusion about the lake's origin. Okeechobee is the largest lake situated solely within the borders of the United States, and, like smaller lakes to its north, it has the characteristic rounded shape of a sinkhole lake. Lake Okeechobee is a relatively new feature of the Florida landscape. It is less than a few meters deep in most places (Kirby et al., 1994). There are several theories of how the lake formed (none of them karst based), the most reasonable being that it is a natural low area surrounded by higher coastal or marine deposits. Certainly karst processes are acting upon the limestone underlying the lake, but the lake itself did not form from a massive collapse of the Winter Park variety.

The New Wales Gypsum Stack Sinkhole

One of the strangest sinkholes to occur in Florida in recent decades formed beneath a large gypsum stack, causing a serious pollution problem in central Florida. Before examining the exact causes of the sinkhole, it is useful to summarize how gypsum stacks come to be. They are created by the phosphate industry as large piles of mine tailings. The central Florida peninsula is home to one of the richest phosphate deposits in the world. The phosphorus ore is taken from the ground through strip mining using large-bucket drag lines. After the ore is processed, the overburden is used to reconstruct the strip-mined landscape. One of the impurities found in the phosphate ore is gypsum. Unfortunately, gypsum contains small amounts of radioactive material, so at present it cannot be used for any constructive purpose. Thus, the gypsum is collected into mountainous (sometimes multi-acre) piles called gypsum stacks. Gypsum stacks typically contain a central reservoir that holds the waste acids left over from processing the phosphorus ore. Gypsum stacks are common features on the rural western Florida landscape and are often the highest topographic features for miles around.

Gypsum stacks are huge and typically contain tens of millions of tons of waste gypsum and tens of millions of gallons of acidic and radioactive wastewater. The weight of the wastewater and the waste gypsum puts a tremendous strain on underlying natural systems. In the past, the stacks were unlined, which allowed acidic wastewater to filter through the gypsum and into the underlying aquifer. It is unknown what specific effect the acidic waters had on the underlying karst system, but the presence of highly acidic water certainly must have increased porosity, at least locally. Of course, there also are concerns about the impact of the radiation that escaped into the subsurface along with associated trace pollutants. In addition, over the long term, natural gypsum deposits can dissolve readily under acidic conditions to create karst landscapes (Paukstys et al. 1999).

On June 27, 1994, an environmental disaster occurred when a sinkhole opened beneath the New Wales gypsum stack owned by IMC-Agrico (figure 3.7). When it happened, the sinkhole was 15 stories deep and 160 feet in diameter within the 200-foot-high gypsum stack. It is believed that the volume of the hole was 2 million cubic feet and that between 4 and 6 million cubic feet of wastewater entered the subsurface through the sinkhole (Satchell 1995). At that particular site, 23,000 tons of byproduct were produced each day (Fuleihan et al. 1997).

The size of the New Wales sinkhole is remarkable. The void space that received the water and gypsum must have been huge. Although voids of this size certainly can exist in the Florida landscape, it must be pointed out that this particular gypsum stack was unlined and that acidic waters filtering into the subsurface could have enhanced any preexisting void underneath the gypsum stack. Engineers working with IMC-Agrico examined the sinkhole through boreholes and via geophysical techniques, and they were able to stabilize the landscape by sealing the void with 3,800 cubic yards of concrete. The highly acidic waters present at the site complicated the work. Nevertheless, the grouting effectively sealed the sinkhole and prevented further leakage into the subsurface.

To prevent such problems in the future, regulators now require extensive geophysical investigation prior to the building of gypsum stacks. In addition, it is required that gypsum stacks be lined in order to prevent leakage into the subsurface. Although the technology has improved sufficiently to help to prevent future problems, gypsum stacks and their associated wastewater remain environmental problems. The hurricane season of 2004 saw three major hurricanes over the phosphate production areas of central Florida. Extreme rainfall and heavy winds caused the breach of a gypsum stack owned by Cargill Crop Nutrition on September 5 during the height of Hurricane Frances in Gibsonton, Florida. Approximately 60–70 million tons of highly acidic water flowed into a creek that drains into Tampa Bay. Certainly some of this water leaked into the subsurface, where it could further enhance karst processes.

Sulphur Springs Sinkhole Group

A complex group of sinkholes is found in central Tampa in the Sulphur Springs neighborhood (figure 5.5). Sulphur Springs is a second-order spring (one with a discharge of 10–100 cubic feet per second) that flows through a short spring run into the Hillsborough River. A major tourist destination in the early twentieth century, the region around the spring is now highly urbanized and part of central Tampa. The Sulphur Springs drainage basin is complicated because although the surface drainage basin is only 7.5 square miles, the area under the spring is cavernous, and it is likely that water drains to the spring from a much greater area. Seventeen sinkholes are found within the spring's drainage basin (Stewart and Mills 1984). According to the Florida Geological Survey, the expansion of the city over the course of the twentieth century has been accompanied by a decline in Sulphur Springs' water quality (Rosenau et al. 1977).

Figure 5.5. Sulphur Springs in Tampa, home to a complex group of sinkholes.

Over the years, extreme rain events caused extensive flooding within the Sulphur Springs drainage basin. This is not particularly surprising, however, as the region can easily receive several inches of rain in a day and it is difficult for the sinkholes to process so much water draining to the subsurface. Flooding was especially troublesome in the northern reaches of the basin, the furthest from the Hillsborough River. To alleviate these problems, storm-water managers built storm-water systems that rapidly brought water directly to the sinkholes. This alleviated some of the flooding problems, but it caused a serious decline in water quality to Sulphur Springs, leading, in 1986, to it being closed to swimming due to the presence of unhealthy amounts of coliform bacteria. Recently, the City of Tampa built a swimming pool on the site using treated water, but the spring remains closed to bathing.

Interestingly, the flow through the sinkholes has changed over the years. Dye tracing conducted in 1958 showed that the springs in the drainage basin were connected (Menke et al. 1961; Stewart and Mills 1984; Trommer 1987). However, heavy flooding in the northern portion of the drainage area in the 1980s indicated that the system was not functioning as an effective drain. A new dye trace was conducted in 1987 by the City of Tampa that indicated that the northernmost sinkhole, Curiosity Sink, was not flowing into the Sulphur Springs drainage system. It is thought that clay dumped into the sink during construction caused the sinkhole to plug. In order to remedy the flooding problems, pumps were installed to drain water from the sink directly to the Hillsborough River (Beck et al., undated). Water is diverted from other sinks as well to prevent the pollution of Sulphur Springs. For example, water-retention ponds have been built near Orchid Spring in order to reduce the impact of polluted storm water on the spring system. In addition, Wallace (1993a, 1993b) notes that a storm-water retention pond associated with I-275 contributes pollution to Sulphur Springs through a sinkhole that opened within the pond. Exploration by cave divers of the conduits leading to Sulphur Springs indicates that the conduit receives water from numerous sources that flow under the city. Also, the conduit flows under the main Interstate (I-275) that serves that part of Tampa.

The research on the sinkholes at and around Sulphur Springs demonstrates the connectivity and interrelationships among sinkholes, groundwater, springs, and human activity. Normal storm-water flow into the springs deleteriously affected the recreational value of the springs for the community. Although a link cannot be conclusively made, the Sulphur Springs neighborhood has declined significantly since the springs closed. Many of the small cottages that were built in the early part of the twentieth century as vacation homes are now part of the broader urban decay of the central portion of Tampa. Although attempts have been made to restore the character of the springs by building a neighborhood swimming pool at the site, the park was closed long enough for the area to lose its recreational character. The water-retention ponds found near the sinkholes have worked quite well in recent years to prevent flooding, although the system failed in 2004 during an extreme rain event associated with Hurricane Frances in early September, when the ponds in the northern portion of the drainage basin near Fowler Avenue and 15th Street breached and flooded the area for several days.

Spring Hill Sinkholes

During the early 2000s, the Tampa Bay area and much of the Florida peninsula experienced a severe drought. Water table levels were at historic lows. Streams like the Withlacoochee disappeared, and wetlands dried up. In the midst of this drought, sinkholes formed throughout the state. With tables lowered, water was no longer present to cushion and stabilize the cover atop underground void space. When rains came, the added weight of the moisture worked to destabilize the cover. Perhaps the best example of drought-induced sinkholes comes to us from Spring Hill, Florida (figure 5.6).

Spring Hill is a small suburban community in southern Pasco County. The area is only a few miles east of the coast and sits in an active sinkhole area located between the Brooksville Ridge and the Gulf of Mexico. In July 2001, the area had experienced well below average rainfall for many months, and water tables were quite low. Then heavy rains hit the area in early July, with four inches of rain falling over a few days. By July 14, more than a dozen sinkholes had formed in one neighborhood. One house suffered serious damage, and there were several problems with road stability. The sinkholes happened suddenly, indicating that they were cover-collapse sinkholes. Luckily, no one was hurt in the incidents, but property damage was severe.

Sinkhole swarms create a sense of community unease. Concern for personal safety as well as for the safety of family and friends coincides with anxiety about property damage. No one wants to lose the home into which so much time and money has been poured over the years. And once a sinkhole swarm forms, an entire neighborhood's property values can come into question. The presence of a sinkhole swarm gives prospective buyers the uneasy

Figure 5.6. Spring Hill in Hernando County, the site of numerous recently formed sinkholes.

feeling that the property occupies a hazard zone. Unfortunately, whether it is legitimate or not, just this perception can diminish the value of a property, even sending an entire community into decline.

At the close of 2001, 65 sinkholes were confirmed in Hernando County, with many of them located in the Spring Hill area. In June 2002, another sinkhole swarm formed in the Spring Hill area, causing further property damage.

Clearly, the community rests on a very active collapse zone, and it is likely that more sinkholes will form in the future.

Another sinkhole cluster formed in June 2002 near the end of a drought. This time, a series of sinkholes opened in the Orlando area. Many of these sinkholes were much larger than the sinkholes reported at Spring Hill. One of the largest occurred at Woodhill Apartments, where a 45-meter-wide and 18-meter-deep sinkhole formed at the edge of two large apartment buildings. The sinkhole drove dozens of residents from their homes and did extensive property damage. Other sinkholes that formed during the same period forced the closure of Interstate 4, the main connector between Tampa, Orlando, and the East Coast.

Summary

Regardless of where they form or their size, sinkholes are fascinating landforms. They are particularly interesting when they directly impact our lives in some way. As we have seen in the case of Paynes Prairie, sinkholes have shaped Florida's history. At the same time, they pose potential risks. When their seemingly random formation damages roads or water mains, sinkholes disrupt our day-to-day activities. They can be heartbreaking when they damage our homes, our property, or us. Knowing that they are common features of our landscape should help us remember that people have been living with sinkholes for generations. Similarly, knowing that physical injury from sinkholes is extremely rare should relieve us of much of our worry that a sinkhole might harm us or our loved ones. As we will see in the next chapter, the threat of sinkhole activity has led scientists and engineers to try to develop systems for predicting sinkhole formation and to better apprise us of the risks.

6

Detection and Mapping

People are naturally afraid of sinkholes. This is partly a result of the uncertainties surrounding sinkhole insurance and sinkhole policy. For a few people, though, this fear borders on "sinkhole phobia." They are terrified the earth will swallow their home whole, sucking loved ones and possessions into the underworld after it. While catastrophic sinkholes do happen from time to time, sometimes with tragic consequences, most sinkholes are far from being Dante-esque events. In fact, it is likely that most sinkholes in Florida go unreported due to the subtle nature of their expression, and when homes *are* affected by sinkholes, the damage is typically not of a magnitude that should induce dread about natural earth processes.

Where are sinkholes located? How are they mapped? These seem straightforward enough questions, yet difficulties are involved with answering both for a number of reasons. First, because the processes leading up to sinkhole formation are located underground—typically as the expansion of void spaces formed from the solution of limestone—detection isn't always possible without special equipment. Similarly, detection of the causes of any particular property damage is costly due to the high-tech nature of the methods employed in the assessment process. Second, sinkhole mapping is problematic because sinkholes' topographic expression varies tremendously. Some sinkholes are quite deep and fall readily within contour intervals used in traditional mapping techniques while many have minimal relief and are thus not detectable using common topographic maps. Another problem with mapping sinkholes is that their size varies a great deal, from very large ones that show up clearly on commonly available maps to very small ones that aren't even identified on maps. Of course, due to the fact that sinkhole formation is a common phenomenon in the state, it is difficult to keep maps up to date. Indeed, as noted above, most sinkholes are probably not reported, and there are benefits to not reporting sinkhole occurrences near or under homes.

Yet there is still a clear need for a better understanding of the location of sinkholes and the voids that cause them. Also, there is keen interest in mapping sinkholes in the state. This chapter reviews how sinkholes are detected in the field, summarizes the way sinkholes are mapped using both traditional mapping techniques and remotely sensed data, and concludes with a summary of the status of sinkhole mapping in Florida.

Sinkhole and Void Detection

Sinkhole and void detection techniques are used as research tools by karst scientists to better understand sinkhole formation and by geotechnical firms to assess claims made by homeowners of sinkhole damage. The techniques include simple field observation to detect voids, ground-penetrating radar, and drilling.

Field Observation

There are a number of potential causes of land subsidence in Florida. For example, buried organic deposits can oxidize to cause the land surface to shrink. This can occur when buried peat deposits oxidize due to the changing nature of aquifer systems. There has been some degree of concern about this phenomenon as regional aquifer levels began to decline in the late 1990s due to the pumping of aquifers and a severe drought. Another cause of subsidence induced by oxidized organic matter results from poor preparation of construction sites. Sometimes in the site preparation process, downed trees, tree stumps, or wood construction debris is left on the sites where homes will be built. When this occurs, the buried organic matter oxidizes over time, leaving a void space into which part of the foundation of the home can slip or collapse. Another cause of land subsidence is the action of shrink-swell clays. These clays, typically within the montmorillonite family, shrink and swell as they get wet and then dry. The mineralogy of these clays makes them suitable for embracing water in the crystalline structure during wet times of year. During dry periods, the water is shed from the mineral structure, causing the clays to decrease in size. Some of these clays can increase or decrease in size significantly. When they are located within an aquifer, there is little problem because the clays are permanently wet and thus of constant size. If the clays lie within a zone of wetting and drying, however, or, if due to regional groundwater withdrawals they become aerated, the ground will subside, particularly if the deposit is thick or regional in extent.

In the field, it can be difficult to discern among the possible causes of land subsidence. In some cases, the evidence takes the form of a clear, round depression that immediately suggests sinkhole formation as the cause. In other cases, however, where a depression is subtle and not easily discernible, causality can be unclear. The causality is especially difficult to determine when the only evidence of possible land subsidence is damage to a home in the form of cracked walls or out-of-alignment windows and doors. Field observations are thus only partially useful in assessment of subsidence events, and for this reason other tools must be used to determine causality, including ground-penetrating radar and drilling. Each of these techniques involves a certain investment in training and equipment, which is why geotechnical firms that specialize in such types of assessment are so important. It must be noted that improvements in the technology available to scientists who utilize this type of equipment is steadily ongoing. For this reason, it is important that anyone interested in obtaining an assessment of potential sinkholes investigate the qualifications of the firms conducting the work.

Ground-Penetrating Radar

Ground-penetrating radar is a geophysical tool that uses particular electromagnetic waves to detect variations in the subsurface. Applications of geophysical techniques to detect subsurface features were first developed in the early 1900s. These approaches were largely used to detect groundwater or buried ore deposits and involved creating continuous flows or pulses of electromagnetic energy in simple arrays (Daniels 1996). A wide variety of applications were developed through the twentieth century, including significant developments during the exploration of the moon in the 1970s. With the advent of desktop computers and concomitant software improvements, the technology advanced rapidly in the late twentieth century. Several important advances have been made in image processing, in airborne deployment of radar, and in accuracy. Such advances in hardware and software technology should enhance the field and better its understanding of the subsurface. Early in the development of ground-penetrating radar, it was evident that a wide variety of subsurface anomalies could be found. These include not only groundwater features and ore deposits but also ice layers, salt deposits, soil variations, archaeological deposits, and, most important for the topic of this book, void spaces.

How is ground-penetrating radar able to find these features? Radar waves are high-frequency electromagnetic waves that are found all around us. Radar is used in a variety of applications, from discerning weather phenomena such as cloud form and height to tracking planes. In geophysical applications, a machine is used to send radar waves into the ground. Once the wave is generated, it enters the ground and the energy is absorbed or reflected off of subsurface anomalies. Ground-penetrating radar applications measure the time between the transmission of an electromagnetic wave into the ground, the reflection of the wave, and the reception of the wave at a surface antenna.

When the energy is pulsed into the ground, it is initially traveling at the speed of light. However, once the waves enter the earth, variations in the subsurface cause the waves to travel at distinctly different speeds. Indeed, subtle changes in sediment type, moisture conditions, porosity, and mineralogy can cause wave velocity modifications. When the waves are reflected off the facies in a sedimentary or rock layer and returned to a receiving antenna, it is relatively easy to process the wave information to map the variations over time. Ground-penetrating radar works best when variations among subsurface materials are great. That is why the technique is so suitable for karst studies. The variation between void space and rock is quite high, and voids and associated raveling zones are easily found through GPR imaging techniques. In most settings, ground-penetrating radar is only useful from a few tens of feet to a few hundred feet in depth, but recent advances in borehole ground penetrating radar (USGS 2000) allow measurement of anomalies to depths of over 3,000 feet (http://www.amcl.ca/geophysicalservices_boreholegpr.html/). Thus GPR usually detects subsurface voids that can cause problems for property owners.

There are several pieces of equipment needed to conduct GPR surveys, including a transmitter, a receiver, a processor, and a display unit. The transmitter creates the radar wave energy that is pulsed into the ground. It is important to have some idea of the target depth, because the maximum depth that can be measured using ground-penetrating techniques is largely dependent upon the frequency of the transmitted radar waves. There are a variety of transmitters on the market that produce particular frequency ranges.

Low-frequency radar in the range of 100–250 MHz is suitable for most applications. However, higher frequency transmitters in the range of 500–800 MHz are appropriate for shallow surveys that require a great deal of resolution (http://www.amcl.ca/geophysicalservices_boreholegpr.html/). Importantly, radar will not work well in high-conductivity materials. Thus no matter what the depth limitations of the equipment, a limiting factor in ground-penetrating radar surveys is the depth to the zone of saturation. Also, materials with high conductance, such as clays and salt, do not allow significant radar penetration, and thus signals are unable to detect significant variations below their surface. Interestingly, unsaturated sands and bedrock provide some of the greatest attenuation rates for radar. Thus in Florida, where so much of the bedrock in the state is covered with a veneer of sand, ground-penetrating radar is an excellent technique in the detection of sinkholes and associated raveling zones. In other parts of the world, geophysical assessment using resistivity is commonly used. Most modern GPR equipment has a combined transmitting and receiving antenna in the same unit, which requires a transducer to toggle back and forth between transmitting and receiving pulses. The pulses are emitted at a rate of between 25,000 and 50,000 pulses per second (Conyers and Goodman 1997).

In order to complete a survey, the antenna is dragged, carried, or flown over the area of interest. Most ground-penetrating radar surveys for sinkhole detection are completed using drag units (figure 6.1). Surveys are typically conducted as discrete transects in a back-and-forth fashion until the area is covered. Global positioning system (GPS) mapping techniques are particularly useful in conjunction with GPR to accurately locate surveys on the planet. The distance between transects is dependent upon the level of accuracy one wishes to obtain in the study. The closer the transect lines, the finer the detail that will emerge between transects. There are a variety of GPR antennas on the market, and it is important to study the characteristics of the units and try them in the field before deciding on the appropriate unit for the application.

Once ground-penetrating radar signals are transmitted and received, they have to be processed in order to make sense of the data. Just as there is a great variety of GPR hardware options to choose from, there is also a variety of signal processing software packages available that interpret the reflected radar pulses. Typically this software requires some set-up parameters to be entered for particular sites, although some software packages are written for particular applications that take into consideration the scale of the project and the type of information needed for the application.

No matter the specific software one utilizes, the processing is conducted using techniques that average the many pulses received per second to obtain reflection data. The pulse data are filtered to remove anomalously high- and low-frequency and long wavelength noise. Once such noise is removed, it is easy to create traces of the digital data set on some type of output device.

Figure 6.1. A GPR system. Photo courtesy of GeoView, Inc.

Modern machines now have real-time screen imaging systems. In contrast, older machines required saving the survey results in some form of digital memory for printout back at the office. New machines have excellent imaging capabilities that are useful for digital communication using the World Wide Web or software programs like Power Point. Similarly, with current print technology, standard and highly enhanced survey images may be printed for distribution in reports, articles, and books (see figure 8.1)

A typical unprocessed survey transect trace will show background noise in the form of horizontal bands that may include a signal from the actual technician working with the antenna or some other non-anomalous feature of the survey area. It is important that such bands be removed prior to interpretation as they can mask important information. However, in some instances, the bands may represent near-surface features of interest, such as buried road surfaces. There are a variety of software options for enhancing ground-penetrating radar transect outputs. Most of these are used to provide cleaner imaging to better depict boundaries between subsurface layers and anomalies. For example, the technique known as deconvolution is used to remove traces from the record that become convoluted during transmission.

Once an image is created through the processing of the GPR information, interpretation of the data can begin. Interpretation of radar logs can be difficult without one's having some degree of experience (1) in the field and (2) with the anomalous condition under investigation. Particularly for homeowners whose properties are being investigated for the absence or presence of subsurface voids or raveling zones, the radar logs can look like random series of squiggly lines that make no sense. Nevertheless, skilled technicians are able to clearly delineate subsurface changes with great accuracy. In central Florida, where Quaternary marine sands over the clayey Hawthorne Formation rest unconformably on top of Cenozoic limestone, the pancake stack stratigraphic sequence appears as even horizontal lines on a radar log. Where sinkholes or raveling zones are present, the horizontal lines become disjointed and the traces become crenulated. Such clear disruptions in the subsurface landscape as depicted in the radar logs can be utilized as evidence of the presence of sinkhole or raveling activity.

The common assumption that sinkhole mapping is a relatively straightforward process, however, is incorrect and based on misconceptions about the expression of sinkhole forms on the landscape. By definition, sinkholes are closed depressions and thus should show up on topographic maps as hatched, closed-depression lines or as water bodies, yet many sinkholes are not large in spatial or geographical extent. In typical U.S. Geological Survey maps of a scale of 1:24,000, only sinkholes greater than 20–40 acres are mapped because of scale issues. Also, they may only show up on maps as designated wetlands or open water if they are not topographically distinct. This is problematic in Florida, where sinkholes may not fall within standard USGS topographic contour intervals of 5 or 10 feet. In fact, many sinkholes may have only a few feet of relief. While these features may be locally important for drainage and may have subtle, but significant, ecological variations due to soil moisture conditions associated with the relief change, they would not be discernible using most standard mapping techniques.

The scale problems involved in sinkhole mapping are compounded in small-scale maps that show a county, region, or the state. At smaller scales, Florida sinkholes commonly disappear from maps, leading one to conclude that a mapped area is a featureless plain devoid of geomorphic variation. Nothing could be further from the truth. Unfortunately, this misperception takes the place of real knowledge of the importance of the subtle karst features present at the surface and their interaction with subsurface groundwater systems.

Such misperception is further supported by modern maps that show developed portions of Florida, where it is difficult to distinguish water-retention ponds from sinkholes. As Florida continues to develop, it will become increasingly difficult to distinguish different kinds of water features from sinkholes on maps. Of course, aerial photos and satellite imagery may be able, at least in part, to circumvent such problems, although they have their own difficulties. The technology of airborne laser swath mapping (ALSM) holds promise as a newly developing technique for mapping surface karst features. Several of these mapping innovations are discussed below.

Map Analysis

Topographic maps of Florida are a suitable starting point in doing any map analysis of sinkholes. The entire state of Florida has been mapped by the U.S. Geological Survey at a 1:24,000 scale. These maps contain an abundance of topographic information at 5- or 10-foot contour intervals. They are readily available in digital or paper format from the mapping agency or from private map companies. They are used by a variety of organizations, from the state water management districts to county and local governments to assist with environmental planning. As noted above, their use in sinkhole analysis is limited, as many sinkholes fall outside of the spatial or topographic scale limitations of the maps, but they provide a regional topographic context for karst features. And in parts of the state where there is significant relief due to karstification, particularly in the Ocala and Brooksville Ridges areas and in portions of the Panhandle, these maps are very useful for mapping sinkholes of topographic significance.

When mapping sinkholes using topographic maps, one must recognize the limitations of using printed contour intervals as a guide to delineation. The contour intervals on the map are a close approximation of the topographic features on the ground and may not be exact. Thus any sinkhole mapping using this technique has errors inherent with the original data set. Nevertheless, it is a useful exercise to complete this type of mapping in order to ascertain characteristics of the karst terrain that are helpful in interpreting geomorphic history and helpful for environmental planning.

To map sinkholes using topographic maps, it is easiest to work with electronic versions of these maps within a Geographic Information System (GIS).

Then one can trace depression contours to delineate possible sinkholes. The benefit of using a GIS to complete this mapping is that it is relatively easy to then calculate a variety of morphometric characteristics of the landscape. Some of these are quite common characteristics, such as sinkhole area and sinkhole density, but more complicated morphometric characteristics—including shape forms, axis orientations, and edge crenulations—may be readily calculated using data tables created through a standard GIS.

The results of morphometric analysis have many applications. Sinkhole area calculations are useful in wetland and water-storage studies, and sinkhole density is used to determine the intensity of karst formation processes. The orientation of axes can be used to estimate regional joint patterns and other structural and lithological characteristics of the bedrock. The shape of sinkholes may be used to ascertain formation history, with coalescing arc-shaped features indicating intersecting sinkholes and thus multiple sinkhole events. Crenulations of contour lines may indicate postformation erosion processes. The benefit of conducting a morphometric analysis of sinkholes is that it provides a mathematical description of the landscape that allows comparison with sinkhole regions in other parts of the world. Morphometric analysis also provides a basis for classification of individual karst regions into specific subregions. In some instances, the analysis is useful for understanding the evolution of the landscape with time.

As noted above, the scale of the map that is used in morphometric analysis has a great deal to do with the outcome of any analysis. A small-scale map, for example, generalizes subtleties in the landscape. This may be useful for broad morphometric analysis, which attempts to characterize the karst features of large areas, such as the entire state or a water-management district. Large-scale maps, however, have a greater potential to more accurately represent sinkhole characteristics. While there is no national or statewide map at scales smaller than 1:24,000, many site-specific maps showing sinkhole distribution have been created by individuals, agencies, and private companies. Unfortunately, there is no statewide attempt to collect these maps and they often exist in the gray literature, which consists of private and public technical reports.

The maps that exist within the gray literature hold a great deal of information that may be of use to karst researchers. As noted earlier in this book, there are a number of centers of karst research in the state of Florida, and many of them have reports on file that are neither widely circulated nor readily available from libraries. Also, there may be a great deal of information available at the county or local government level, where region- or site-specific

Table 6.1. Morphometric analysis types

Type of Morphometric Analysis	Calculation	Use
Area, relief, and volume	Calculated using area function of GIS and imported topographic information	Useful, along with relief information, to calculate water storage and volume
Sinkhole density	Area of sinkholes/area of map	Useful for measuring intensity of karstification
Axes orientations	Long primary axis is the longest line that can be drawn across a sinkhole. The secondary axis is the longest line that can be drawn perpendicular to the primary axis. The orientations are calculated in degrees from north.	Useful for determining regional structural control to sinkhole formation
Axes ratios	Secondary axis length/primary axis length	When the axis ratio approaches 1, the sinkhole approaches a circular form caused by a single event. Axes ratios that approach zero indicate a more linear, and thus more joint controlled, sinkhole. Low numbers may also indicate complex sinkholes.
Crenulations	The number of irregular features on a closed depression contour line	The greater the number of crenulations, the more irregular the sinkhole form. There are two reasons why the crenulations may be high: multiple sinkhole event or postformation erosion.

Note: This table displays the kinds of morphometric analyses that can be conducted on sinkholes.

studies are conducted for specific applications. Many of these studies may have been conducted under the auspices of an engineering or environmental department of local government. Recent reports are often available electronically or through the Internet. Of course, a great deal of the gray literature is proprietary, meaning that the information contained within the report is owned privately and not available to the public. Each year, hundreds of sinkhole reports are investigated by geotechnical firms in order to confirm or deny the presence of sinkholes under structures, and these reports are not available to the public due to the nature of the information: few homeowners wish it to be known that there is a confirmed sinkhole on their property. However, on occasion, these reports make their way into the public record, often through court cases, and thus are useful from a scientific perspective. Although it is tempting to urge broad public disclosure of the results of sinkhole investigations by geotechnical firms, it does not seem in the best interest of the general public to release this information for several reasons. First, investigations are paid for privately and thus disclosure would breach the rights of ownership. Second, while some people argue, for the sake of public safety, that it would be useful to know where particular sinkhole-damaged homes are located, because sinkholes are located in broad areas in Florida, disclosure could damage property values for tens of thousands of homeowners who live in sinkhole-prone regions but whose homes are not at risk. Third, disclosure of the private information damages the rights of individual property owners who wish to maintain the value of their property by having their homes repaired and the sinkholes mitigated. Finally, the scientific benefits of disclosing sinkhole reports do not outweigh the damage caused by the release of the reports, as the scientific information is site-specific and does not provide a systematic approach to understanding the distribution or the causes of sinkhole formation. Nevertheless, it is enlightening to examine reports that make it into the public record. They provide interesting analyses of data collected in the field from ground-penetrating radar and, often, coring. Although no one to date has systematically gathered information from the reports in the public record, it would be a useful exercise to (1) examine the data for patterns that would help us better understand sinkhole distribution and formation and (2) assess the reports for quality and consistency.

Most modern mapping is conducted using some sort of GIS, an electronic map attached to a digital database. The digital database provides the thematic elements of the map, and these thematic elements may be easily manipulated statistically in order to create discernable patterns. The digital database may be updated as conditions arise, making a GIS approach to mapping very flexible in regard to the rapid growth of scientific information. Indeed, many believe that digital mapping is one of the most important technological achievements of this generation, and GIS is fast becoming a standard mapping system that is part of the modern technological revolution. This technology, along with the integration of global positioning system techniques, comprises the standard mapping tools now used by field scientists engaged in sinkhole research (Gao et al. 2006). In fact, these technologies are now available on the Internet, on cell phones, in cars, and in numerous other applications that range from the trivial (cell phones) to the scientifically significant (satellite mapping of Mars).

It would be beneficial to develop a statewide, or perhaps even a nationwide, GIS database of sinkholes. Such a database could incorporate information from a variety of sources, such as topographic maps, aerial photos, satellite imagery, and the gray literature. Also, the GIS system could be updated to add new sinkholes as they occur. What makes GIS so powerful is that it is easy to search the database for anything. For example, it would be easy to search for sinkholes of a certain depth or sinkholes in a given county. One of the key elements of any GIS is data management and control. It is important in any GIS exercise to have data quality-control measures in place to make certain that sinkholes that become part of any database are verifiable and appropriately defined.

Data quality in GIS poses a significant problem. GIS database information can be organized into two broad categories: (1) data generated prior to the technological advances of computerized mapping and GPS and (2) data generated after the development of these modern technologies. The data generated prior to recent technological advances is significant and includes important base data such as topographic and locational information imported from USGS topographic maps. Also, spatial information developed prior to these advances may be found in the gray literature, in technical reports, and in peer-reviewed studies published in scholarly journals. In contrast, more recent information has the benefit of precision mapping when compared with earlier spatial information. The quality discrepancies presented in these two types of information lead to some degree of tension in the application of mapping in sinkhole studies. Base maps widely in use may not be as accurate as they could be. Also, accurate spatial information must sometimes be "nudged" to fit less accurate base maps.

The problem can be compounded because Geographic Information Systems are integrated into many applications. For example, sinkholes mapped using USGS topographic maps may be imported into a more accurate county-level base map that has attributes mapped precisely using GPS technology. The imported sinkhole data may not synch spatially with other mapped features in the county. In such situations, the GIS technician must either accept the data as inaccurate but nonetheless useful or find some way to edit the less accurate data to more appropriately locate sinkholes. Whatever the approach, qualitatively discrepant databases have repercussions when they are made widely available for multiple applications. For example, GIS databases may be shared with public agencies working on wetlands or engineering applications. In such situations, sinkhole location information, often taken to be absolutely accurate, proves otherwise, especially when ground truthing of the system involves the use of older spatial information imported into GIS.

Certainly these older databases are extremely useful when imported into a GIS, but it is important to recognize the problems inherent in these systems when conducting any type of analysis or when proceeding with site-specific studies.

The widespread integration of GIS into public records at the federal, state, and local levels makes the situation complex. It is often difficult to evaluate the quality of GIS databases without having quality metadata associated with the system. The metadata associated with GIS products provide details on the creation of the spatial information and typically include information on the creation of the base map and the grid and map projection used in the base map. Also included is information on the creation of the spatial coverages. It is important that anyone creating a GIS database on sinkholes include a detailed metadata file to assist future users in evaluating and updating the information. Without a good metadata file, it will be extremely difficult for future researchers to effectively use the database. As technology advances in digital mapping, it is anticipated that more standardization and uniformity will occur that will greatly assist those involved in sinkhole mapping.

Discussion of GIS isn't complete without consideration of spatial information included in the gray literature. As noted above, there are a variety of approaches used in conducting geotechnical investigations of sinkholes. Therefore, the integration of a number of different studies of sinkholes into a single GIS system is difficult. Indeed, there can be problems in generalizing this information into a database. How can diverse types of investigations be put into a uniform system? This is where the construction of Geographic Information Systems becomes so significant. The developer of a GIS must keep different types of reports and approaches distinct. This enables the user of the system to query the database by type of report or by methodology.

Yet in the end, once a detailed database has been created and made available, the genie is out of the bottle, so to speak. There is great danger that sinkhole spatial databases may be misused. For example, it would be relatively easy to combine all sinkhole information collected from a variety of sources into one map, but this map would be inappropriate for most uses due to the mixed methodology of data collection. However, such a map could be powerful in the hands of the general public in that it could be used for insurance redlining or might inappropriately influence property values. Some individuals and groups are reluctant about developing a statewide database due to the problems of quality and the potential for misuse. Thus, if an effort to create a sinkhole GIS proceeds, it is crucial that a nonbiased organization take on the task of developing a statewide GIS mapping of sinkholes using standard criteria and a peer-reviewed process. It is important that whoever takes on the effort be regarded by all parties as nonbiased in order to resolve the current mistrust among stakeholders on sinkhole issues in the state. As sinkholes have become a big business in the state of Florida, spatial information regarding sinkhole occurrences can be used in legal cases.

The development of a statewide sinkhole database is not a new idea. In the 1980s, the Florida Sinkhole Research Institute, developed at the University of Central Florida under the direction of Barry Beck, began creation of a statewide database of reported sinkholes. The institute, started at the beginning of the technological revolution of digital mapping, recorded location information for reported sinkholes. However, at the time, many sinkholes were unreported and the database was seen as lacking a systematic approach. Nevertheless, the institute did a great service in the advancement of the knowledge of sinkhole formation and distribution in Florida. When it closed in the 1990s, its database was transferred to the Florida Geological Survey. As we will see later, the FGS is active in enhancing the state sinkhole database, but the work of this organization is limited in scope and there is thus a need for a statewide effort to create a more comprehensive GIS sinkhole database.

Aerial Photo and Satellite Analysis

Photographic images taken from airplanes or satellites can be used to assist with mapping sinkholes (Dinger et al. 2007) (figure 6.2). As with traditional mapping using some type of base map or GIS, this type of remotely sensed information for sinkhole mapping has pluses and minuses associated with the technology.

Using photographic images to map sinkholes is especially useful for sinkholes that show distinctive moisture variations within the depression. Moisture conditions can range from open water features to the presence of hydric soils and associated wetlands vegetation. Depending upon the resolution of the photograph and the time of year, these moisture conditions are easy to discern as distinct circular features on the landscape. Open water is easy to spot. Hydric soils, usually signaled by dark coloration, are also easy to map in Florida. Dry soils, by contrast, usually have the light coloration typical of the quartz sand that covers much of the state. Variations in vegetation are also excellent indications of the presence of sinkholes that can be mapped as coinciding with ecotonal edges of circular features on the landscape. When

Figure 6.2. Satellite image showing numerous karstic lakes in Pinellas County and a portion of Hillsborough County. Aerial photo courtesy of Cindy Shaw.

open water is present, it is important not to map sinkholes as associated with the edge of the water but to examine the landward extent of hydric soil conditions. Variations in topography also can be helpful in the delineation of sinkholes in these images. Similarly, there are useful tools that assist with topographic analysis of aerial photos and satellite imagery.

There are several drawbacks associated with aerial photographs and satellite images. Perhaps the most important of these is the difficulty of consistently placing images within standard map projections. In the case of aerial photos, the images are taken from airplanes that vary in altitude and pitch. Although modern aerial photo specialists can remove these discrepancies, many aerial photos also have resolution problems. Often, some type of software is needed to "rubber sheet" the aerial photos within the selected map grid. For example, when using standard GIS or aerial photo image-processing software, known points on a base map can be linked electronically with the digital aerial photo scanned into the mapping software. Once these known points are set, the aerial photo is stretched, or rubber sheeted, to fit the base map. When sinkhole delineation is conducted on the aerial photo images, small location errors can creep into the database due to slight variations caused by distortion during the rubber sheeting process.

Another problem associated with both aerial and satellite imagery is that they capture a distinct moment of time that may or may not be environmentally standard. In other words, moisture or vegetation conditions may not be the same now as when the image was collected. So although one may use these moisture or vegetation parameters when mapping sinkholes, the actual edge of the depression may not be located at the boundaries of the discernible parameters due to seasonal or annual variations. It is important, therefore, to understand the climate conditions present when the image was collected. Was it a dry or wet year? Was it collected in the dry season or wet season? What were the antecedent meteorological conditions prior to the collection time of the image? All of this information is readily available for Florida through the National Weather Service.

Another challenge in the mapping of sinkholes using aerial or satellite imagery is their potential partial obscuration by cloud cover or shadow. If a photo is taken on a summer afternoon, there is likely to be some degree of cloud cover due to the daily formation of cumulus clouds in Florida. On aerial photos, the presence of clouds creates shadows that make delineation of subtle sinkhole forms difficult. On satellite images, some areas may be fully obscured due to cloud cover. These challenges can be mitigated by taking care to select images collected in conditions that were as cloud free as possible. The high number of aerial photo and satellite images available on the market usually means there are several options available to researchers.

Due to Florida's rapid urbanization, many of the attributes one would use to map sinkholes using aerial or satellite images are continually being altered. For example, moisture conditions often are modified by land development. Sinkholes can be filled. Water-retention ponds required to collect drainage from impervious surfaces can mimic sinkholes on the landscape. Although many created ponds are rectangular, some are circular and have open water and associated soil moisture variability common to natural sinkholes. Thus in areas where development has progressed rapidly, the use of aerial photos and satellite imagery to map sinkholes can be limited due to the complicated nature of the urban landscape.

Of course, another challenge presented by aerial photos is their identity as artifacts of a particular time period: obviously, they will not show newly formed depressions. While this may not be a problem in the short term, the technique's limitations where developing a sinkhole database is concerned—particularly a database covering a rapidly urbanizing area in active karst terrain—must be noted. Ground surface is complicated by the fact that urbanization destroys subtle topographic variations caused by sinkhole formation. Thus, modern aerial photos and satellite imagery have limitations that may preclude their use in highly urbanized or urbanizing areas. The use of historic aerial photos, however, holds a great deal of promise for advancing better understanding of the distribution of sinkholes in the state prior to widespread development. Modern aerial photos and satellite imagery are best used in undeveloped portions of Florida, where human alteration of sinkhole formations is unlikely. In heavily developed areas of the state, the photos are useful when used in conjunction with other data sets that allow the technician to confirm the presence of sinkholes and other topographic features.

Airborne Laser Swath Mapping

Airborne laser swath mapping (ALSM) is a relatively new technique that allows detailed mapping of topographic features on the earth's surface. The technique is a form of remote sensing that provides extremely accurate depictions of the landscape. A wide variety of applications for ALSM exist, including the mapping of sinkholes (Seale et al. 2007; Vacher et al. 2007).

The use of ALSM to map topography is based on the evolution of light image data recovery (LIDAR). LIDAR is a technique that was developed in the field of atmospheric science to map cloud cover and form. The idea behind LIDAR is that light beams, or laser beams, are emitted to a reflective body. Because the speed of light is constant, the return time of the reflected light beam provides a distance from the source of the light to the reflective object. Thus, by shooting laser beams at clouds, scientists are able to measure the spatial features of clouds with great accuracy. In the last decade, the technique has been used to ascertain topographic features on the planet.

The LIDAR technique was refined over several years to allow detailed mapping of the ground surface using equipment flown on airplanes. The new technique, ALSM, uses a light transmitter and a receiver flown over the target areas. Multiple beams are transmitted from and received at the plane to measure variations in the topographic landscape. It is important to note that the instrumentation uses state-of-the-art global positioning techniques to accurately place the airplane and its equipment within three-dimensional coordinates in the air. This allows relatively easy calculations of distances between the airplane and the ground surface using the known speed of light constant. Because multiple sites are selected for distance measurements owing to the swathlike approach used in the technique, hundreds of thousands of measurements are made in order to do a simple topographic mapping of a small area. Indeed, the technique requires a sophisticated computing setup in order to handle the large data sets generated.

Once the data are generated, they must go through some form of filtration largely dependent upon the purposes of the mapping. Interestingly, ALSM maps all topographic features, natural and unnatural, so that trees, telephone poles, and buildings all are visible on the generated maps. If one is interested in seeing these features on the landscape, there is very little filtering that must be done. However, in landform analysis and in most topographic mapping, these features must be removed in order to understand the lay of the land. There are many approaches that can be used to remove unwanted features. For example, trees are relatively easy to remove, because the ALSM detects not only the trees but also the land surface between the limbs or leaves of the trees. So a simple subtractive algorithm can eliminate these features. The same can be said of telephone poles and other point features on the landscape. Buildings are much more difficult to eliminate due to their size and dimensions, but it may not be essential to remove buildings in order to assess the karst landscape using ALSM maps.

Once filtration is complete, a variety of ALSM maps can be created that depict the ground surface. A variety of programs can be used to display ALSM data. The one most commonly used is Surfer, which allows three-dimensional display of geographic data. It is quite easy to import ALSM data into this software package, or into other suitable programs, to display the landscape. A number of different types of maps can be created using these programs, including standard topographic maps and shaded relief maps.

Because of their great accuracy in measuring all landscape features, ALSM maps have a distinctive appearance. Indeed, the technique, which allows high-resolution, detailed topographic mapping with an accuracy to within a few centimeters or inches, provides a stunning look at our altered planet. In addition to displaying the natural landscape, ALSM maps clearly depict alterations created by road building, construction, and other earth-moving activities. Clearly this technique holds great promise for a variety of future applications. In Florida, where the topography is subtle, the technique could be very useful for managing storm water, for understanding coastline change, and, of course, for understanding our karst terrain.

In karst studies, ALSM holds potential for mapping sinkholes and other depressions that fall outside of standard contour intervals. As noted elsewhere, standard contour maps created by the USGS for Florida have 5- to 10-foot contour intervals, which limit their ability to define sinkholes or other depressions that do not have local relief that is less than the interval. Certainly many karstic features are not identifiable on maps due to this limitation. In addition, the limited contour interval does not allow for mapping the accurate shape of the karstic features. Anyone working with USGS topographic maps must use depression contour lines as guides to the form of the sinkholes. However, with ALSM, we are able to depict these depressions with a great deal of accuracy. In fact, we are able to see that the karst landscape is much more complex than depicted on topographic maps. We are truly at a revolutionary moment in understanding Florida karst due to improvements in mapping and imaging technology. No longer are we reliant only upon partly accurate topographic maps. We can now more fully understand the development of karst due to ALSM technology.

On the other hand, ALSM is expensive, partly because it involves developing technology. Only recently have ALSM maps of large areas become available for public use. The newness of the technique and the variety of imaging processing approaches used to remove things like trees from the data set make standardization of the technology, and thus the maps, problematic. At this time, then, more standardization needs to be completed prior to the development of a uniform system of ALSM mapping. Nevertheless, ALSM has great utility for local mapping.

Another challenge presented by ALSM mapping is that the topographic details depicted confront the user with a range of information never before encountered in traditional topographic mapping. In standard USGS topographic mapping, a great many generalized topographic features appear on the maps. However, with ALSM mapping, all topographic information is present. This creates two broad challenges. First, it often requires the user to develop some scheme of generalization in order to complete analysis of the information. Take, for instance, a situation in which one wishes to map depressions. With ALSM, because all depressions are shown on a map regardless of whether they have a relief of a few centimeters or a few meters, new ways of deciding generalized limits must be created for better analysis. Another challenge involves data management and interpretation. In the last century, a great deal of research on morphometry and landscape interpretation was done using standard USGS topographic maps in conjunction with aerial photos. The new details provided by ALSM challenge conventional approaches to map analysis in these fields. For example, an archaeologist working on finding archaeological sites associated with sinkholes may have used a topographic map showing depressions as a base for the association. Now, with ALSM topographic mapping, new approaches to interpreting the archaeological information must be created to better understand the relationship between humans and the landscape.

Discussion of the challenges inherent in ALSM isn't complete without mentioning the difficulty of distinguishing natural from humanly created features. Although the ALSM data clearly identify anthropogenic features such as roads and housing developments, the many subtleties depicted make interpretation of the causes of those features difficult, particularly in developed areas. The causes of subtle depressions, for instance, are many and include natural landscape formation and a variety of human activities including construction of water-retention areas and borrow pits. Even so, the details provided by ALSM make research into our modern landscape infinitely more interesting and rich.

It must be reiterated that ALSM is a new and revolutionary technique in topographic mapping that in combination with rapid developments in image processing provides researchers with wonderful opportunities for advancing science. Although there are concerns with the use of ALSM, the technology is providing exciting possibilities for new research.

Historic Records

The use of historical records in sinkhole mapping is important in Florida because a large proportion of the state's landscape has been altered in some way through human activity. There are three types of historical records useful in researching the location of pre-development sinkholes in a region: historic maps, historic aerial photos, and archival records.

Historical maps provide an interesting but limited amount of information on pre-development sinkholes. There are a number of different types of historical maps. Old state or regional maps may provide some limited information, although any sinkholes shown on these maps are likely to continue to exist due to their substantive size. Of more use are more local maps, and of these, there are a number of different types that can be helpful. It is important to note that the pace of development in Florida is rapid. The population grows significantly each year, and with this striking increase, land development too occurs at an accelerated rate. Maps that are not particularly "historic" often predate urbanization or suburbanization and thus provide information about the predevelopment landscape. Old local government maps as well as maps and plans created by private organizations can thus be useful to our visualizing the karst landscape before development. With improvements in geographic information systems, historic maps can even be placed in layers within a digital database to better elucidate the karst system and the role of urbanization in the modification of the landscape.

Old aerial photos are excellent sources of historical information. Many parts of Florida have aerial photo coverage dating back to the 1920s. Many of these photographs are of dubious quality and, because some early photographs have been lost or destroyed, coverage in certain parts of the state is incomplete. However, extant images easily can be placed within a geographic information system to compare the historic to the present landscape. Historic aerial photos are particularly helpful in mapping where urbanization activities filled in or destroyed existing sinkholes. This information is useful not only for geomorphic analysis but also for tracking wetland and habitat loss due to the unique hydrologic conditions present in many sinkholes. Again, with development progressing rapidly in the state, aerial photos significantly more recent than the 1920s are useful in mapping karst features.

These more recent photos typically have a better resolution and are often in color.

A final source of historical material worthy of discussion here is archival information. This may take the form of technical reports, databases, or newspaper articles. A wide variety of technical reports on karst topics are produced each year by public and private agencies. The utility of these reports in describing or mapping sinkholes varies, as does their quality. Nevertheless, I recommend searching local libraries and agencies to gather as much pertinent material as possible to support the conduct of sinkhole research. There are also a number of sinkhole databases. Perhaps the most well-known is the one created by the now-defunct Florida Sinkhole Research Institute, which was housed at the University of Central Florida. This database, now maintained by the Florida Geological Survey, contains locations of reported sinkholes in the state (FGS 2011). County-level databases also record reported sinkholes. These databases, often housed in engineering departments, typically include only those sinkholes that did some type of property damage. Proprietary sinkhole databases are owned by consulting firms or insurance agencies but are unavailable to the public. Finally, archival information on sinkholes can be found in newspaper articles. Such articles typically feature local sinkholes that either did damage or are in some way unique. Most Florida newspapers have searchable databases that can prove useful in assessing historical sinkhole occurrences.

A word of caution: because they only record reported sinkholes, databases and newspaper archives are of limited use in a systematic analysis of sinkhole activity in Florida. Hundreds of sinkhole events each year go unreported. Likewise, newspapers only report dramatic events apt to elicit broad public interest. Thus one must use such information with great restraint and with an understanding of its limits. Such information would better serve assessments of historic property damage in the state rather than it would the rate of natural sinkhole occurrence. As development expands in Florida, especially into the highly active karst areas of Orlando and Tampa, the reporting of sinkhole events will increase.

The Status of Sinkhole Mapping in Florida

The location of sinkholes interests a great many people and organizations, particularly within the scientific community. There is understandable concern, however, that the reporting of sinkhole locations may lead to devaluation of properties or to insurance redlining. Regardless of such concerns, there have been and continue to be attempts to extensively map sinkholes in the state. This section reviews some of these attempts and provides a context for understanding their results.

As noted earlier in this chapter, scale issues are problematic to sinkhole mapping. Because sinkholes exist at vastly different areal sizes and depths, the selection of a mapping scale necessarily biases the outcome of the mapping exercise to exclude particular sinkhole sizes. For example, a comparison of depression contours on 1:24,000 and 1:100,000 maps of the same area of Florida would present very different overviews of the karst terrain. The 1:24,000 map would show significantly more detail than the 1:100,000 map. The latter map would show only the coarsest karst features and would certainly not show many of the smaller features represented on the 1:24,000 map. Refining the 1:24,000 map to a 1:1000 map or larger would certainly allow for significantly more detail. Thus the choice of base maps is crucial and greatly influences the results of any mapping exercise.

Another problem inherent in sinkhole mapping is defining what is meant by the term "sinkhole." For example, one researcher might choose to map only circular sinkholes while another might map only uvalas, or coalescing sinkholes, within a single depression. The differences are subtle, but they have a significant impact on the outcome of the mapping exercise and the eventual interpretation of the data. In addition, researchers might decide to map only reported sinkholes. In such cases, some sinkholes might not be expressed as depressions at all but show up in structural damage or in subsurface investigations. And because the phrase "reported sinkholes" indicates some level of pre-selection prior to inclusion on the map, the definition of a reported sinkhole is crucial to the understanding of the databases based on them. County engineering departments often keep a record of reported sinkholes, though the requirements and compliance for reporting varies from local government to local government, and the quality of these databases must be analyzed prior to interpretation. So not only is the choice of base maps important, but the database that one uses or creates greatly influences the eventual results. No approach will be perfect as maps will always generalize the reality of the ground surface. But it is important to note that there is a great deal of selection and generalization that enters into any mapping exercise, and the choices that one makes may exclude sinkhole data.

Is there such a thing as a perfect sinkhole map? There are certainly limitations to any mapping exercise. However, there is a great deal of opportunity to create multiple sinkhole databases that allow broader interpretation of sinkhole form and distribution using geographic information systems. Indeed, it is useful to map sinkholes at different scales using different criteria for a variety of purposes. It would be useful to consolidate this information within one database so researchers could better understand the distribution of sinkholes across our landscape.

The first statewide attempt at understanding the distribution of sinkholes was conducted by William C. Sinclair and J. W. Stewart, who in 1985 published a map titled "Sinkhole Type, Development, and Distribution in Florida." The map was prepared by the U.S. Geological Survey in cooperation with the Florida Department of Environmental Regulation and the Florida Department of Natural Resources, and it may be seen online at the Florida Geological Survey's website at www.dep.state.fl.us/geology/geologictopics/sinkholedevelopment.htm/. The map divides the state into four sinkhole areas defined by the thickness and type of cover above limestone. Area I, classed as bare or thinly covered limestone, is found in three distinct areas of the state: the southern portion of Florida, including all or portions of Monroe, Dade, Collier, Broward, Palm Beach, Hendry, and Lee counties; the Big Bend area of Florida, extending in an arcuate band from Wakulla County to Hillsborough County; and a small area in the north-central Panhandle in Jackson, Washington, and Holmes Counties. The map authors define these areas as having few generally shallow and broad sinkholes dominated by solution sinkholes. They note that these sinkholes develop gradually.

The cover in Area II is 30–200 feet thick and is defined as incohesive and permeable sand. The sinkholes here are reportedly few in number, shallow, and of small diameter and gradual formation. In these locations, cover-subsidence sinkholes dominate. The sinkholes are found largely in the eastern peninsula, extending from extreme northeastern Dade County to Volusia County in the north, and in small areas in west-central Florida, largely along the Brooksville Ridge. Area III, defined as having a clayey, low-permeability cover 30–200 feet thick, is in scattered locations largely in the northern half of the state. The area includes large portions of Pinellas, Hillsborough, Pasco, Polk, Lake, and Orange Counties in west-central Florida; large areas of Flagler, Putnam, Marion, Alachua, and Union Counties in northeast Florida, and significant areas in the eastern and central Panhandle region. Here, according to the authors, sinkholes are the most numerous of any areas in the state. Also, sinkholes here are of varying size, develop rapidly, and are of the cover-collapse variety. Finally, Area IV is classified as having a cover of more than 200 feet. The sediments consist mainly of cohesive sediments with discontinuous interlayered carbonate beds. Here, there are very few sinkholes, although large deep sinkholes do occur and are of the cover-collapse type.

Although not a sinkhole map per se, the Sinclair and Stewart map provides a framework for understanding the nature of sinkholes geographically across the state. The map is largely causal in that the authors state that the controlling factor in sinkhole occurrence is the thickness and the type of land cover.

The first comprehensive attempt at developing a sinkhole database was conducted by the Florida Sinkhole Research Institute in the early 1980s under the direction of Barry Beck. Prior to that time, dating to 1907, the Florida Geological Survey kept a record of reported sinkholes and other karst information. However, the Sinkhole Research Institute expanded on those efforts. Over 1,700 sinkholes were mapped from reported occurrences throughout the state. The result of this effort was a database that includes information on location, size, form, and occurrence. Unfortunately, this effort was not maintained, and it now serves mainly as an important document that helps us understand the distribution of reported sinkholes over a particular time period that ended in the early 1990s, when the funding for the institute was removed. It is also important to note that the database includes only those sinkholes that were in some way reported. Sinkholes that occurred in undeveloped areas were likely underreported. The FGS currently maintains the database (one can report a sinkhole to the FGS on its website).

There have also been several efforts made to map sinkholes at the regional level in order to better understand their formation. For example, Ann B. Tihansky of the U.S. Geological Survey in Tampa completed a detailed study of sinkholes in west-central Florida (Tihansky 1999) in which she describes several scenarios that provide triggering mechanisms for sinkhole formation and demonstrates through mapping that sinkhole distribution is partially controlled by the nature of the cover. This interpretation is consistent with the work of Sinclair and Stewart (1985) described above. Another researcher, Dean Whitman of Florida International University, and his colleagues have mapped sinkholes in a portion of central Florida with an eye to better understanding the nature of sinkholes and water levels (1999). Whitman's work used high-resolution digital topographic data to construct maps. Several other researchers have conducted site-specific studies in various portions of the state.

Insurance companies and geotechnical firms also produce private sinkhole maps. These maps are proprietary and thus unavailable to the general public, but they help insurance companies better understand the distribution of sinkholes and their financial risks. These companies utilize the same techniques others use to map sinkholes, although it is likely that the maps and associated databases differ due to the particular needs of these organizations.

7

Sinkhole Policy

Sinkholes are vexing features on the modern landscape (Zhou and Beck 2007). When they form, they can cause considerable damage to our property and infrastructure, leaving us with damages, repairs, and a sense of unease about our firmament (Quinlan 2007). In June 2001, Raymond Kelly of Lutz lost his sense of comfort about the stability of his property. After a loud crash, he found that his home was severely damaged: doors ruptured and jammed and part of the roof torn away. But what caused the problem? Kelly immediately thought the problem was a sinkhole because several other homes in his neighborhood had experienced similar problems. His insurance company hired a geotechnical investigator who did an assessment of his property using standard investigative techniques and found that the problem was not a sinkhole but settling sand and buried organic matter. Kelly was shocked with this assessment and found that his homeowner's insurance covered sinkhole damage but not damage from settling sand or buried organic deposits. As a result of the investigator's findings, Kelly hired an attorney to challenge the insurance company's evaluation. The geotechnical company the attorney hired to evaluate the property found a different cause than did the insurance company, concluding that the damage was caused by sinkholes. How can two geotechnical companies come up with distinctly different results?

My insurance company recently dropped sinkhole coverage from my home, built in 1961 in Temple Terrace. If I wanted coverage, I needed to request it. When I did, I was told that I needed to pay $50 for an evaluation. I set up an appointment for the evaluation and the person wandered through the property and my house. Sinkhole coverage was denied. I was stunned. There was absolutely no explanation from the insurance company, just a general denial saying that the evaluation I paid for showed evidence of risk. But they had produced no evidence for the risk.

The home in Seffner, Florida, in which a man died in 2013 after a sinkhole formed and a portion of the house collapsed had undergone a similar inspection, but the home passed that inspection. Why are sinkholes covered in some policies while other subsurface irregularities are not covered? This chapter reviews current sinkhole insurance rules and some of the problems that arise because of rules that currently exist. In addition, it will discuss several other policy issues related to sinkholes, including the use of sinkholes for water retention. Given that sinkholes are often associated with springs and caves, a brief summary of current regulations regarding springs and caves in Florida is also included.

Sinkholes and Insurance

Most claims to insurance companies never get to the point of litigation. Indeed, the example presented above, Raymond Kelly's house, is rare. Insurance companies pay many claims each year and, when challenged, they often choose to settle with homeowners rather than incur the expense of a legal challenge. But how did we get to the point that we are arguing subsurface subtleties in a court of law?

Sinkhole coverage was first mandated in Florida in 1969. At the time, coverage was optional, although insurers were legislatively obligated to offer it (Salomone 1986). While many opted to get a sinkhole rider on their insurance, many property owners did not choose the rider and thus incurred losses when their property was damaged by sinkhole formation. In 1979, the legislature expanded the sinkhole insurance requirement by mandating sinkhole coverage for all homeowners. The language of the state statute regarding sinkhole insurance (State Statute 627.706) was as follows:

1. Every insurer authorized to transact property insurance in this state shall make available coverage for insurable sinkhole losses on any structure, including contents of personal property contained therein, to the extent provided in the form to which the sinkhole coverage attaches.
2. "Loss" means structural damage to the building. Contents coverage shall apply only if there is structural damage to the building.
3. "Sinkhole loss" means actual physical damage to the property covered arising out of or caused by sudden settlement or collapse of the earth supporting such property only when such settlement or collapse results from subterranean voids created by the action of water on limestone or similar rock formation.

4. Every insurer authorized to transact property insurance in this state shall make a proper filing with the office for the purpose of extending the appropriate forms of property insurance to include coverage for insurable sinkhole losses.

The law was a significant development in the protection of property owners in the state of Florida. There are several interesting aspects of this law (Salomone 1986). It clearly states that every insurer must provide sinkhole coverage for property insurance and that the insurance covers not only the structure but also all of the property within the structure. However, the law does not include the land on which the property rests. This is not unusual, as common homeowners' insurance does not cover the land on which a home rests. Thus if some catastrophic event happened to the land, the insurance company would not be responsible for paying out any damages, but if the structure and its contents are damaged through the damage to the land, the structure and the contents are covered.

Another important aspect of the sinkhole rule resides in the phrase "sudden settlement or collapse." These words caused a great deal of debate when the law was enacted. As we have seen, sinkholes can form rapidly or very slowly over time. If your home were on top of an area that saw rapid sinkhole formation, your situation would certainly fit the definition of "sudden settlement or collapse." These types of sinkholes form mainly in zones where cover-collapse sinkhole formation dominates. However, what if your home were in an area where there was slow raveling of sand into a subterranean void in a zone of Florida where cover-subsidence sinkholes dominate?

This question was resolved in the case of *Zimmer v. Aetna Insurance Company* (1980), which involved a property damaged from a sinkhole that was not catastrophic but formed slowly. Aetna Insurance denied the owner's claim because they believed that the sinkhole formed slowly and did not fit the requirement of "sudden settlement or collapse." Zimmer sued Aetna in court and lost. The trial court agreed with Aetna's interpretation. Zimmer appealed to the Fifth District Court of Appeals, and this time, the court sided with Zimmer and clarified the statute. They specifically interpreted the "sudden settlement or collapse" phrase for the future, and all sinkhole insurance claims after this court case were impacted by the interpretation.

The district court particularly took issue with the insurance company's interpretation of "sudden" and "settlement." The word "sudden" was interpreted by many in the insurance industry to mean that the sinkhole formed rapidly. In contrast, the court chose to interpret "sudden" as "coming or occurring unexpectedly, unforeseen or unprepared for." In this definition, the term loses its time element and is seen much more as an unexpected event of some undefined duration. Also, the insurance companies interpreted settlement to mean a true and visible loss of ground volume where the sinkhole is clearly visible on the land surface. Again, the circuit court disagreed with this interpretation and defined settlement as "a gradual sinking of the structure, either by the yielding of the ground under the foundation or by the compression of the joints or material."

These interpretations had major implications for both insurance companies and for property owners. Instead of having a clear definition of sinkholes as entities that are topographic features on the earth's surface, a sinkhole was defined as a subtle feature that causes some form of land instability that may or may not have a surface expression. This distinction is crucial, because the legal definition of a sinkhole is quite different from a geomorphic or geologic definition of a sinkhole. The distinction is largely for pragmatic reasons. Certainly there are homes that are damaged by karst processes that lead to the formation of geomorphic sinkholes. I should note, however, that a bit of ground settling as a result of raveling does not constitute the formation of a geomorphic sinkhole. This nuance is important because there is so much discord on this issue in the courts of law and among those in the geologic professional community. Certainly there is no argument from the insurance industry when land over a large subterranean void collapses and causes structural damage to a home. These features are clearly sinkholes. But when cracking foundations or some other subtle property damage occurs, there can be multiple causes of the damage and evaluations of the subsurface are required. Because the circuit court imposed a definition of "sinkhole" that is not a geomorphic definition, there are often subtleties of interpretation by experts in the field, often pitting one expert against another in legal cases. The state modified the sinkhole insurance rules and now requires coverage for catastrophic subsidence, but it makes standard sinkhole insurance optional.

As noted above, in Florida there can be multiple causes of property damage that mimic the damage caused by circuit court–defined sinkholes. Some of the more common causes are shrink-swell clays, the oxidation of organic matter, and poor construction or site preparation. Shrink-swell clays are common sedimentary deposits around the world. In fact, the USDA, in the classification of soils of the United States, defines one of the 12 major soil classification units as being dominated by shrink-swell clays. This soil is quite common in Texas and Louisiana, although it is found in patches throughout the United States. One of the properties of this soil is that as it expands and contracts, large cracks form in the soil that can be tens of centimeters wide and meters deep. When these types of sediments are buried, as they often are in Florida, they can cause the ground to shrink during the dry season and heave during the wet season. The reason for this is that the sediments can take in water within their crystalline structure, causing an expansion of volume many times that of the dry mineral.

In Florida, the location of many of these clay deposits near the surface is known. The deposits often occur in regional patches as a unit within the Hawthorne Formation, but the clays also may be quite localized and impact small areas. Thus the presence of shrink-swell clays must be considered when evaluating subtle structural damage. Because Florida goes through distinct rainfall seasonality, the water table of the surficial aquifer and the Floridan Aquifer often varies over the course of a year or during extreme wet and dry conditions. If the water tables intersect shrink-swell clay deposits, the land may pitch and heave regularly, causing regional disruptions of the ground surface. If water tables in a zone of shrink-swell clays are lowered as a result of pumping, the contraction of clays may occur and cause ground surface instability.

Similar problems exist in areas of buried organic matter. However, unlike clays that shrink and swell with changing water conditions, organic matter oxidizes when aerated to cause shrinkage of the ground surface. Organic matter may be found buried as distinct peat layers beneath the surface. Peat deposits are quite common in Florida because marine transgressive and regressive sedimentary sequences were deposited across the land surface of the state. Some of the peat deposits are meters thick and provide clues to the environmental history of the state. Like the clay deposits, they occur both as predictable units that can be mapped, and they occur as localized patches in the subsurface. Many of the peat deposits are saturated beneath the surface. Under these conditions, the peat is a stable geological unit, but when peat is exposed to oxygen through the lowering of water tables, the peat can oxidize and cause the land to subside to cause surface instability. Also, if in the preparation of a building site some large organic object, such as a tree trunk, is buried, the void left after the object decomposes can cause instability leading to foundation problems. The presence of subsurface organic matter can therefore mimic the expression of subtle circuit court–defined sinkholes.

Finally, it must be noted that poor construction practices can also mimic damage caused by circuit court–defined sinkholes. On many occasions, poor site preparation, poor foundation construction, and poor building construction have been found to be the actual causes of property damage. Any insurance claim for subtle property damage will thus require an examination of the overall quality of the property construction.

Clearly, the evaluation of any property claim for sinkhole damage is not a very easy task. There are multiple factors that can cause structural damage to a home, and each of the subterranean causes of ground instability can have exactly the same surface expression. In addition, there must be subsurface investigations in order to assess the actual cause of damage.

Compounding the problem is the fact that in order for a sinkhole damage claim to be granted, the presence of a sinkhole underneath the property must be proven.

> "Sinkhole activity" means settlement or systematic weakening of the earth supporting such property only when such settlement or systematic weakening results from movement or raveling of soils, sediments, or rock materials into subterranean voids created by the effect of water on limestone or similar rock formation. (Florida Statute 627-706)

This has been interpreted in courts to mean that the presence of a sinkhole beneath the property must be proven. Void space and/or an associated raveling zone must be found through subsurface investigations. This is easily done by test borings and by ground-penetrating radar. However, testing is not flawless, and the raveling zone may exist off site. For example, if a sinkhole forms beneath Property A, the raveling zone can impact a broad area beyond that property (figure 7.1). Structures on Property B may be damaged by the sinkhole. Unfortunately, as interpreted, the structures on Property B would not be covered by homeowners' insurance because the sinkhole did not occur under the property. Indeed, normal subsurface investigation would not discover a sinkhole under Property B, precisely because such investigations are limited to the delineation of the actual homeowner's property, in this case Property A.

This leads to an interesting question with regard to sinkhole liability. If a sinkhole opens beneath one property and affects other properties, is the owner of the property on which the sinkhole forms liable for damage to other properties? To date, this issue has not been explored in courts, although with the growing attention to sinkhole policy in Florida, it is only a matter of time

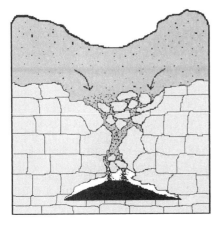

Figure 7.1. Raveling involves the slow downward movement of sand and other sediment into a void space beneath the surface.

before it enters the public discourse. Liability issues have been tentatively examined in relation to human-induced sinkholes with little resolution. For example, when sinkholes form in areas of regional water drawdown, there have been attempts to blame the formation on the owner of the well (often a local government). Because of the multiple causes of sinkhole formation, this has not met with success. Indeed, although a sinkhole may be induced by human activity, the void could have already existed before the advent of the human action, and the sinkhole would have formed at some unknown time in the future anyway. In such cases, it is difficult to assess losses due to human action.

Currently, insurers are only required to cover damage from catastrophic ground collapse. Homeowners must pay for extra insurance for sinkhole coverage—if they can pass an inspection. It is the inspection process that is particularly vexing.

The Growth of the Sinkhole Industry

Since the adoption of the sinkhole insurance requirement, there has been astonishing growth in what can reasonably be called the sinkhole industry. Arrays of different types of businesses have developed in the last three decades that are a direct reaction to the state rules. These businesses fall largely into three categories: geotechnical firms, sinkhole repair specialists, and real estate speculators. As with any industry impacted by state regulation, each of these businesses has a distinct interest in state sinkhole policy and is active in either maintaining the status quo or changing the policy to better suit their needs and/or the needs of the public.

Geotechnical Firms

A number of geotechnical firms specialize in evaluating sinkhole insurance claims. In some instances, the companies were founded solely to evaluate claims, and in other cases, established companies expanded their expertise in the 1980s in response to the sinkhole law. The evaluation of sinkholes based upon the state's definition of a sinkhole for insurance purposes was not uniformly conducted initially after the development of the law. A variety of techniques were utilized to identify sinkholes in the subsurface, and the quality of the evaluations varied considerably from company to company. In evaluating how sinkhole claims are handled by insurance companies, the Florida Geological Survey (1997) reported that a geotechnical firm will first conduct shallow borings to find subsurface voids if there is not a clear surface expression of a sinkhole. When shallow borings do not identify any type of cause for the structural failure, deep borings are conducted. In west-central Florida, ground-penetrating radar is always used to try to identify any subsurface anomalies. In other areas of the state, GPR is sometimes used to locate the best place to bore deeply into the earth to try to find subsurface anomalies.

Regardless of the techniques used, geotechnical firms sometimes find themselves interpreting similar information in different ways, with one firm representing the homeowner claiming that a sinkhole caused the damage and one firm representing the insurance company claiming that the damage was not caused by a sinkhole. Because of these problems, a "sinkhole standards summit" was held in 1992 to try to develop "certainty and consensus to the issue of minimum standards to be employed in the determination of the [sic] whether a sinkhole has caused damage to a structure" (FGS 1997, V-7–V-8). The summit brought together the movers and shakers in the sinkhole evaluation field and included the director of the Florida Geological Survey, professional geologists and engineers in the consulting area, academics, and a representative from state government. The meeting was the first (and only) organized public meeting that brought together a wide variety of professionals to discuss the scientific problems associated with the insurance rule and to address inconsistencies in consultants' sinkhole evaluations. The meeting resulted in several findings, summarized below, from the FGS (1997, V-9–V-18):

1. A structural engineer, a geotechnical engineer, and a geologist should ideally evaluate all claims. However, due to cost considerations, at a minimum one licensed professional with experience in sinkhole claims should conduct the evaluation.

2. It is impossible to standardize sinkhole investigations. However, a process that includes pre-site assessment, on-site assessment, and detailed site assessment should be conducted. The pre-site assessment should include interviews with the property owners and other parties with information on the claim, information from the literature or other sources, and the development of theories of the cause of structural damage. The on-site assessment should include an initial site visit that assesses the local geology, hydrology, structure, and structural damage. The detailed site assessment should include subsurface evaluations including borings and geophysical analysis. At this point the theory of formation should be evaluated based on data produced.
3. A sinkhole research institute should be formed to help researchers and industry share information and keep up with the latest sinkhole science.
4. Training of insurance adjusters on sinkholes and sinkhole claims would assist the industry.
5. The reports produced by geotechnical firms are often complex and difficult to understand. Thus, clear summaries must be provided for property owners.
6. An evaluation of the "structural adequacy" should be part of the evaluation.
7. Building codes should be improved to prevent sinkhole damage. The report stresses that the Southern Building Code Congress International has a "deemed to comply" standard that could reduce the number of homes damaged by sinkholes. In addition, the group noted that most sinkhole damage is to areas that have less strict building codes than commercial buildings that rarely have sinkhole damage.
8. Soil tests should be conducted at construction sites in order to determine if shrink-swell clays are present at the site. Foundations can be built to handle the site, but it adds a few thousand dollars to the cost of construction.

These agreements provided a framework for better understanding the problems of sinkhole investigations in the state of Florida, but they did not eliminate the problem of variable results from independent consulting firms. Zisman (2001) states the problem quite clearly: "The author has reviewed reports from some investigators who have refused to acknowledge sinkhole presence if evidence of raveling is not found during the investigations, while other investigators have declared sinkhole presence based on very few confirming sinkhole features" (Zisman 2001, 31). Unfortunately, to the public it appears that some consulting firms serve as paid advocates for the interests of insurance companies and some consulting firms serve as paid advocates for the interests of home owners. While this may be the case with some firms, most are quite reputable and provide honest appraisals. Yet the perception remains, and the entire industry is tainted by it.

Zisman (2001) describes three detailed methods for assessing an area that would unify the processes used by geotechnical firms in the Tampa region. The assessments are complex and include evaluations of subsurface characteristics through borehole investigations and geophysical soundings in order to determine if the damage was caused by sinkhole formation, organic matter oxidation, shrink-swell clays, or some combination of these processes. Zisman believes that the results of the evaluations could be mapped in order to promote better understanding of new problems across the state.

Although Zisman's work demonstrates that distinct methodologies can be employed to systematically evaluate properties for sinkhole damage, his approach is not widely accepted and there is no distinct approach that is agreed upon by professionals in the field. However, any geotechnical firm's evaluation must hold up in a court of law, and experts on the opposing side would challenge poor evaluative practices. Thus there are checks and balances on the processes that are utilized in the field, and most firms are reputable and use appropriate procedures for evaluating sinkholes. The interpretation of the results is typically the point of disagreement.

The problem with interpretation largely exists due to the difference of opinion as to what constitutes a void space underground. Does the presence of a void space indicate the presence of a sinkhole? Does raveling, or the settling of sand, indicate the presence of a sinkhole? Often the answers are a matter of professional interpretation. Some professionals are quite liberal when defining sinkholes, using the state statute, while others are quite conservative in their interpretation. Those doing the interpretation examine the same data but come to different conclusions based on the interpretation of that data.

There are those who argue that the interpretation depends upon who is paying the professional. They believe that those who evaluate claims for insurance companies take a conservative approach, while those who evaluate claims for homeowners take a more liberal approach. Research has not been conducted to test this hypothesis. There are certainly companies that work primarily with insurance companies and those that work mainly with homeowners, so it would be easy to test this hypothesis to evaluate if there are differences in the interpretation of the rule.

Because of the differences of opinion, some cases end up in court. When this happens, there are significant expenses to insurance companies and homeowners. Sometimes, the insurance companies choose to settle rather than face the costs of litigation. All of these difficulties demonstrate that there are distinct issues with the way the state definition is written. It would be extremely useful if the state definition were evaluated to settle the issue once and for all. Specifically, it would be useful if the state defined what constitutes a sinkhole and what evidence must be present in order for a property to be covered by sinkhole insurance. In addition, it would be useful if the State Insurance Commission provided definitive guidelines for evaluating sinkhole claims.

Of course, what is needed to solve the problem is a complete change in the sinkhole rule. Given all of the problems, it would help to reexamine the entire issue of sinkhole insurance in the state. It seems that there are two radical solutions to the problem: (1) either expand property insurance to include other causes of structural damage such as organic matter oxidation and the presence of shrink-swell clays or (2) eliminate sinkhole coverage altogether. Let's examine the first option. If the insurance industry were required to include other forms of structural damage, the cost of homeowners' insurance would certainly increase. Given the many problems associated with homeowners' insurance in the state due to recent hurricane damage, it is unlikely that insurance companies or homeowners would find this additional cost palatable (though this approach should not be barred from the discussion until a cost analysis has been completed). Let's now examine the second option. If the state eliminated the requirement for insurance companies to cover property for sinkhole damage, insurance costs would go down. While sinkholes can occur anywhere in the state, there are certain areas where they are much more likely. Thus, insurance companies could require riders for sinkhole coverage in particular areas just as they do for flood insurance. Sinkhole science has gotten sophisticated enough that risk maps can be created to assist in assessing areas where sinkhole coverage should be mandated. However, it must be noted again that sinkholes can occur anywhere; moving in this direction would result in a total loss of property were sinkholes to form in areas where insurance is not required.

Clearly neither of these options is particularly attractive. What makes the issue vexing is that it is so scientifically and socially complex. Indeed, the problem rests squarely within the nexus of science and public policy—an area that is neither well researched nor well understood. While eliminating or expanding insurance policies in the state would be difficult, it certainly would be useful to conduct more research on the problems associated with the current sinkhole policy. Due to the public outcry over sinkhole insurance problems, it is likely that the state legislature will take up the issue in the coming years and that sinkhole policy may be radically different from the current situation.

The Eastman Study

Eastman et al. (1995) examined the impact of the sinkhole insurance requirement in Florida. They conducted a survey of the main insurers in Florida in order to assess the number of sinkhole claims paid in the state from 1981, when the coverage was mandated, to 1995 and found that the number of claims expanded greatly over their study period. In 1987, only 35 claims were made, but in 1991, 426 claims were made. The number of denied claims over the time period also increased. In 1987, about one-fourth of all claims were denied, but in 1991, more than half of the claims were denied. The authors of the study provide two possible explanations for this increase. First, more homeowners may be making claims on a variety of damages that may or may not be caused by sinkholes as they have become more aware of the sinkhole issue. Second, insurance companies may be more rigorously working to deny claims. The authors also note that there is no empirical evidence to support either explanation. Yet the increase of claims has certainly affected the insurance industry, and the increased denial of claims has affected the public's perception of the insurance industry.

The authors of the report also studied the distribution of sinkhole claims and found that most of them were in the Tampa Bay region. This result further supports the case that sinkhole risk is not evenly divided within the state. Many who have full sinkhole coverage are not at great risk, while others who have a higher risk pay the same rate as others without risk.

The researchers also examined the value of claims made by homeowners and found that 73 percent of all claims were below $40,000 and 45 percent of all claims were below $20,000. However, they also found that there are many cases in which claims exceed $90,000, indicating that the properties were

Table 7.1. Costs of sinkhole-related losses by year, 1987–1991

Year	Loss and Adjustment Expenses
1987	$691,805
1988	$1,315,450
1989	$1,782,040
1990	$4,539,667
1991	$8,831,031

Note: Costs of losses as a result of sinkhole loss payments to the insurance companies studied by Eastman et al. (1995).

probably total losses. Thus there is a bi-modal distribution to sinkhole payouts with most being below $40,000 and with another peak above $90,000.

The costs of handling sinkhole claims have increased greatly, particularly the costs of handling denied claims. In 1987, the cost for handling a denied claim was $64. By 1991, this cost rose to $1,457. In contrast, the cost for handling a paid claim in 1987 was $2,149 and rose to $2,620 in 1999. Clearly, managing denied claims has gotten more expensive for the insurance industry. Associated with the increased number of claims and the increased cost of handing claims is the increased loss to the insurance industry for sinkhole claims. The authors found that for the companies surveyed, the losses rose from $691,805 in 1987 to $8,831,083 in 1991 (table 7.1). While this is a dramatic increase, the researchers also found that sinkhole losses account for 0.5 percent of the earned premium.

Eastman et al. (1995) also studied the reasons claims are denied, finding that the major reasons for denial are soil settlement, clay shrinkage, oxidation or compaction of organic matter, and bad construction. These are common problems that often appear to be sinkhole damage. Unfortunately, the damage to property of these homeowners was not covered by their insurance and they had to pay for any repair costs or sell the home to a real estate speculator.

Insurance Cancellations and Other Conundrums

Insurance companies cannot nonrenew due to some type of sinkhole claim that is not worth the total cost of the home. Recently, however, the state-run insurer of last resort, Citizens Insurance, changed its rules so that once a home has a claim for the total damage, Citizens can no longer insure it.

This will be a difficult policy to enforce as the claim will then have to somehow be recorded for the life of the property. Will the information be part of a deed? Or will a new level of record keeping be needed to enforce this rule? In addition, it is likely that the rules will be changed in the future to count multiple smaller claims toward the total value of the home, so that if a homeowner makes several small claims due to sinkhole damage, once the claims reach the value of the house, the house would no longer be insurable.

One frequent complaint is that companies refuse to write policies in particular areas where sinkholes are common. It is against the law to nonrenew a particular home due to sinkhole damage, but a company can nonrenew several homes in an area or in entire neighborhoods or counties for any reason. Thus if a company is accruing some type of financial loss in an area, they can cancel policies in that area. The state-run insurance agency, Citizens, is often the insurer for sinkhole-prone regions. Because of all of the complaints from residents about insurance cancellations, the state legislature has shown an interest in looking at this issue in some detail.

One of the other problems is the lack of oversight extended to how homes are repaired after sinkhole claims are paid. Currently, once a sinkhole claim is paid out due to property damage, homeowners can use the money however they see fit. They can do a patch job on the repairs and sell the home as is and make a profit on the claim money and the money from the sale of the home. Insurers and many in the state are interested in requiring homeowners to repair the home as per the specifications of a geotechnical report, thus eliminating non-approved repairs that may plague the home for the life of the structure. Insurers could end up paying out for multiple repairs from a single event if the property is not repaired appropriately immediately after the formation of the sinkhole.

Also problematic has been the argument among some insurance companies that, because homeowner's policies protect the house and not the land, insurers should not be responsible for stabilizing the subsurface. Insurers typically don't object to repairing homes but contend that the ground stabilization process (which will be discussed in the next chapter) should be paid by homeowners as they are responsible for damage to the land. This argument has received a great deal of attention in the insurance industry, for if the insurance companies were to prevail here, it would greatly impact the way sinkhole coverage is handled in Florida. So far, the companies pursuing this strategy have lost in court, making insurers responsible for the cost of ground stabilization when claims are made.

Lack of Development Rules

Appropriate planning and construction could eliminate many of the sinkhole problems in Florida. Unfortunately, there are no rules that delineate areas where development should not occur or that mandate particular investigations due to the likelihood of sinkhole occurrences. Other parts of the country have developed planning rules for karst regions that protect the environment and help to eliminate structural problems. For example, Lexington, Kentucky, developed a sinkhole ordinance in 1995. The Lexington area is underlain by a complex karst system that includes caverns and sinkholes. Many problems, including groundwater pollution and land instability, have plagued the area for decades, and the ordinance proved crucial to directly addressing the problems associated with the Kentucky karst region. In addition, developers in the region were concerned that they could be sued by homeowners with damaged homes in areas where sinkholes are known to occur (Dinger and Rebmann 1986).

In order to develop a parcel in the Lexington area, a developer must map all sinkholes on the property. The map must include the sinkhole drainage area. The developer must then show nondevelopable areas where sinkholes are present. This area must include sinkholes present on the surface today as well as filled sinkholes. Nondevelopment areas are those places on the land where sinkholes are present. Structures or parking areas may not be built on these sites. The local planning commission may require an extension of the nonbuildable area beyond the border of the sinkhole limits if subsurface investigations warrant. Where sinkhole clusters occur, the Division of Engineering may require a detailed geotechnical investigation in order to develop the land between individual sinkholes and in order to assess the ground stability. In addition, the rule addresses storm-water flow in sinkhole areas. Developers can build in sinkhole drainage basins, but they may not allow storm water off the development to flow into a sinkhole. They must provide alternative drainage that doesn't negatively impact another sinkhole drainage basin or stream system.

The Lexington sinkhole ordinance greatly improved the ability of local government to regulate development on its karst landscape in order to protect groundwater and in order to protect the public from poor site planning. The ordinance served as a model for many other communities in sinkhole-prone areas, and today dozens of communities have some type of sinkhole rule. Yet in Florida, these types of rules are not common.

The most significant regulation in the state is found in Hernando County. Hernando County Ordinance 94-8 was adopted to protect the groundwater reserves in the region from pollution. The rule is far-reaching and sets development guidelines designed to provide groundwater protection areas. Several types of situations are defined in the rule, including protection of sinkhole recharge areas. In order to be protected, the sinkhole must be identified as having a direct connection to the aquifer. If there is a connection, the sinkhole is protected from development and a 500-foot buffer surrounding the sinkhole is set aside as nondevelopable. However, sinkholes filled with material of permeability similar to or less than the surrounding material are not protected. It is important to note that the rule was developed solely for groundwater protection. It was not intended to provide guidelines for site assessment for real estate development.

It is interesting to compare and contrast the motivations behind the Lexington, Kentucky, ordinance and the Hernando County, Florida, ordinance. In Lexington, the entire community, including developers, was behind the ordinance that aimed at protecting not only the environment but also the property owners. In Hernando County, the ordinance focused largely on the protection of a regional water resource. One important difference between the two communities is that Lexington does not require sinkhole insurance. Thus when properties are damaged, the only recourse that homeowners have is to sue the developer for poor site selection and building. In contrast, most homeowners in Hernando County have access to insurance as mandated by state law and developers are not held responsible for damages to homes or for poor site selection. So while there may be a need for sinkhole rules to protect homeowners, there is little broad support because any site analysis to investigate sinkholes will cause costs to rise.

However, many communities have set standards for development in karst areas and delineate limestone regions where special rules for development must be met (Limestone Resource Committee 1993). This would be difficult in Florida, as nearly the entire state is vulnerable to sinkhole formation. Yet, as noted, particular regions of the state, specifically the Tampa Bay area, are more prone to sinkhole formation and associated property damage, and it may be useful, therefore, for some type of regional rule in the area and in other more vulnerable areas.

One recent case that demonstrates the need for better subsurface investigations in the state prior to construction is the collapse of portions of the Lee Roy Selmon Expressway in Hillsborough County. The existing expressway extends from south Tampa northeast to downtown Tampa and then to the growing bedroom community of Brandon. Plans for expansion of the expressway, which was well under way in July 2004, included the building of an elevated span above the existing roadway to greatly increase the ability of the road to handle the growing traffic volume. The elevated span was to be held aloft by a series of dozens of single piers (figure 7.2). Suddenly, on the morning of July 6, one of the piers sank 20 feet into the subsurface, causing a collapse of the roadway under construction onto the ground surface. Two workers were injured in the collapse. An investigation of other piers found other problems, including another pier that had started to sink.

The pier problem was blamed on a number of different causes, from improper construction to poor site analysis. At the time of this writing, it is not entirely clear what specifically caused the failure of the pier. It could have been a sinkhole or it could have been compaction of soft sediments beneath the surface. Regardless, the failure delayed the multi-million-dollar project,

Figure 7.2. The collapse of part of the elevated portion of Tampa's Lee Roy Selmon Expressway in 2004 was reportedly caused by a sinkhole. Photo courtesy of the *Tampa Tribune*.

and it is likely that the parties involved will end up in some type of litigation over the cost of the repairs. In addition, the public's confidence in the ability of the government to manage the complex project was shaken and the relationships between the state and its contractors were damaged. What could have prevented this disaster? Certainly there are techniques available to investigate the subsurface where construction takes place. There is no doubt that the local community and the state will examine the requirements for site assessment. There are examples in the state of sinkholes forming beneath roadways, and the risk to expressways is just as real as the risk to homes. Much of the concern for sinkholes associated with roadways in Florida has focused on the potential for groundwater contamination due to storm-water runoff from roadways or the potential for a hazardous waste spill (Padgett 1993).

Often, when sinkholes occur, the property owner tries to affix blame on someone else. That is what happened with the collapse of a portion of the Lee Roy Selmon Expressway. This is not all that different from the fingers that get pointed when homes are damaged due to some subsurface irregularity. It is unlikely that further regulations will eliminate structural damage and failure, but I would not be surprised, given all of the attention given to the expressway failure and home damage in recent years, if new building guidelines are implemented to include more thorough site analysis prior to construction. It is likely that these regulations will increase the cost of construction in the short term, but the potential for long-term savings is high.

No one can argue that there are risks in building any structure in a karst area. But what are the best ways to manage karst lands? Approximately 19 percent of the United States is susceptible to ground instability due to karst processes. Because of this, Kemmerly (1993) developed a multiphase approach that planners can use in assessing karst systems for development potential and for creating zoning rules. The first phase involves delineating sinkhole-prone areas and areas where flooding can be problematic for developments. This effort is accomplished by mapping areas where sinkholes are located and areas that are known to flood due to a lack of a surficial drainage network. The second phase involves field monitoring. In this phase Kemmerly recommends developing a system to map the location of reported sinkholes and conducting a field reconnaissance after extreme rain events to map the distribution of flooding. When heavy rains fall over karst areas, the sinkholes often are unable to effectively drain the excess water and flooding occurs. Knowing the geographic extent of flooding can help to eliminate particular areas from development in order to prevent future problems.

The third phase of the assessment involves the development of a derivative map that puts together all of the information collected in the first and second phases. The map should include flooding potential as well as collapse risk. Kemmerly (1993) suggests three risk categories: high, associated with areas where sinkhole collapse occurs and where there is the potential for new sinkholes; moderate, associated with the presence of sinkholes where joint patterns and moisture conditions indicate a likelihood of sinkhole collapse; and low, associated with areas where sinkholes are present without being associated with joint or moisture patterns. The selection of criteria for risk assessment maps is full of difficulty. While Kemmerly's (1993) criteria are one way of examining risk, there are certainly others that might be more appropriate in Florida. Also, it is likely that the best way to develop a risk assessment map may vary from place to place due to different causes of sinkhole formation and the fact that the subsurface geology and hydrology vary spatially.

Kemmerly's fourth phase, the final one, is an implementation phase and requires the development of sinkhole rules. The rules may require the development of a separate authority or the addition of authority to an already existing governmental body, such as a water management district. Rules related to offsets or site-development requirements should be established, and postdevelopment drainage requirements should take into consideration the karstic nature of the area.

Another approach in developing sound land-use practices in karst terrain was developed by Veni (1999). It is much more detailed than that developed by Kemmerly, involving a complex decision-making plan based upon detailed geomorphic and GIS analysis. There are two broad phases to Veni's approach: data collection and data analysis. Veni developed his model while working on the karst system associated with the Edwards Aquifer in Texas. The first part of phase one is the evaluation of existing karst databases for the region of interest. In some cases, the database may be missing or incomplete, and if that is the case, the database will have to be expanded through field analysis (Veni does not mention this, but map and aerial photo analysis should be conducted as well as field analysis at this stage). Veni suggests that a field survey be done by walking the area with a crew of 3–5 people 10–30 meters apart in order to discover any karst features. Once found, the crews should note the hydrology, lithology, flora and fauna, morphology, sediment present, geologic structure, local geomorphology, and any other notable features. This data should be transferred into a spatial data base for future analysis. Once the karst systems are mapped, they should be geologically evaluated. For example, caves can be divided into types of caves based on origin. The evaluation should include an evaluation of the flooding potential of the region and the potential for sinkhole formation. In addition, more detailed groundwater flow assessments can be completed to better understand how surface water moves through the aquifer within the area of study. Understanding the joint patterns and other structural elements, along with understanding the lithologic variations, further enhances the understanding of the karst system. When this work is complete, Veni (1999) believes that the final phase of the data collection phase is the development of an evolutionary model for karst development. The model will place the karst system within a continuum of karst landscape formation to better understand what the future may hold for the area.

When the model is complete, data analysis can begin in order to develop a land-use model for the region. The first phase in data analysis is determining the limiting factors for development in the area. Often, these are predetermined by government agencies and include limitations for hydrologic or archaeological reasons. Once this is determined, critical areas must be defined within the karst area. The critical areas may be defined by particular karst hydrology, ecology, or uniqueness. The development pressures exerted on the area may also define the critical area. Next, the karst carrying capacity must be determined by evaluating the water resources in the region in order to assess the sustainable water quality and quantity for the area. Such calculations are tricky and are typically based on changing per capita water consumption and average annual rainfall. Because these calculations do not take into consideration extreme drought or flood conditions, one must be careful not to overestimate the carrying capacity of any area.

The next part of the data analysis phase is to limit impervious cover to 15 percent of the land cover. The 15 percent figure is based upon research by Schueler (1994), who notes that more than 10–20 percent land cover in drainage basins negatively impacts the environment. Thus when karst drainage basins have less than 15 percent impervious cover, land development may proceed within planning regulations. When more than 15 percent impervious cover is present, land development should refocus not on developing new lands but on redevelopment of existing impervious land. Once this assessment is complete, the next phase is the analysis of the protection of karst areas. Do current practices effectively protect hydrology, water quality, and other important aspects of the karst landscape? Finally, once all of this information has been obtained, the last phase is the development of karst resource protection zones. Protection can be quite broad and may include land-use restrictions around particular areas, or it may be more site-specific and involve the protection of individual features by engineering berms around sinkholes or cave entrances.

Veni's (1999) approach is based largely on the goal of protecting karst resources from the negative impact of development and the goal of protecting the hydrologic systems in karst regions. In contrast, Kemmerly's (1993) goal is to develop a plan for managing lands based upon the specific hazard and risk to the human population. A combined approach that takes into account the risk from sinkhole formation, the protection of water resources, and the protection of karst systems seems to be the most appropriate way to develop a method for assessing land-use decisions in karst regions. Veni appropriately suggests that a detailed analysis of the subsurface geology and hydrology is important in understanding how karst systems work. Further, Kemmerly appropriately suggests long-term monitoring and mapping of flooding and newly formed sinkholes in order to assist with detailed decision making. Both approaches are useful.

Interestingly, the state of Florida is making efforts on both fronts. The Florida Geological Survey has taken over the management of the sinkhole database started by the Sinkhole Research Institute at the University of Central Florida. Certainly not all sinkholes that occur in Florida are part of the database, but this important effort is one way we can understand sinkhole risk in the state. In addition, the Florida Geological Survey continues to study sinkholes and other karst systems in the state in order to better understand how and why sinkholes form.

While Veni (1999) and Kemmerly (1993) provide helpful guidelines for assessing karst lands, no such assessment is required in Florida. However, local land managers do have the right to impose specific regulations for land use in their districts. As we have seen, some areas, such as Hernando County, have opted to establish specific criteria for development of karst lands, though most karst areas in the United States do not have such regulations. Ultimately, the homeowner takes a risk when buying any property and the homeowner or their insurance company must assume any losses.

Sinkholes and Real Estate

One interesting phenomenon that has resulted from the awareness of sinkhole-damaged homes is the growth in the number of real estate speculators who specialize in sinkhole-damaged homes and properties. In many areas of Florida, one regularly sees billboards or other forms of outdoor advertising for these types of companies. When a house is damaged by a sinkhole, its value decreases. Sometimes, whole neighborhoods become known as sinkhole-prone areas, causing a regional property value decline. Often, individuals whose homes have been damaged by a sinkhole wish to move out either because they do not feel safe or because they do not wish to go through the process of home repair should the damage occur again.

I recently drove through a middle-class neighborhood where it was known that many homes were damaged in some way by sinkholes. Perhaps only one out of every 20 or 30 homes had some type of damage, though some of that damage was quite obvious. Walls were cracked and foundations were undergoing repair. The homes were in overall good shape, but I could see that many of the homeowners had patched cracks in walls or driveways or covered block with some type of facing such as fake brick. Many homes were for sale and the neighborhood appeared to be in some decline. I didn't study the neighborhood well enough to determine the cause of the decline, but certainly the presence of numerous sinkhole-damaged homes had an impact on the property values in the neighborhood.

There are many paths that a homeowner can take when a sinkhole damages a home. It is important to note that not all homeowners have insurance. There was a recent report of an elderly woman whose home was severely damaged by a sinkhole who lost everything. She owned her home outright and did not have insurance. The home was a total loss, and she had to walk away from the property. However, many homeowners in Florida do have insurance that includes sinkhole coverage. If there is some type of damage from what the homeowner believes is a sinkhole, the homeowner must call the insurance company to make a report. The insurance company will then evaluate the claim by conducting a site assessment. If the assessment determines that the damage was not caused by a sinkhole, the homeowner has two options. They can challenge the assessment and hire an attorney to help them fight the insurance company or they can agree with the assessment of the insurance company and pay for repairs on their own.

Property owners have a range of options when managing their repairs. Unfortunately, because most opt to complete just what is needed and do not address the source of the problem, damage to homes can occur repeatedly. In addition, many property owners do not have sufficient money to conduct significant home repairs. They must either take out a loan or complete patchwork that stabilizes the home but does not fully repair the damage. As a result, the denial of a sinkhole claim typically causes deterioration in property value because property owners are not fully able to manage the needed repairs. Often, they sell to real estate speculators at a reduced cost to get out of the headache of having to repair the home.

If an insurance company grants a sinkhole claim made by a property owner, the owner has some options as to what to do with the money. First of all, the owner may take the check cut by the insurance company and use that money to repair the home. As an alternative, the owner may keep the check and sell the home to a real estate speculator as an as-is home. Whether the speculator purchases the property as a result of a denied or granted claim, they are able to conduct repairs to increase the value of home due to the economy of scale. They can hire contractors to work on multiple homes to conduct repairs at a lower cost than could a single homeowner paying for repairs to one home. These types of companies provide useful services to homeowners by giving them the ability to get out of a damaged home at a reasonable price.

It is important to mention that insurance companies, although required to include collapse coverage in homeowners insurance, do have the right to not insure in particular areas. They can pull out of a particular region where sinkhole claims may cause significant losses. Therefore, property owners in entire neighborhoods may have difficulty obtaining homeowners insurance or the optional sinkhole coverage.

Sinkholes and Water

In Florida, there are a number of federal, state, and local rules that help protect our water resources. Given that water is integrally associated with our karst landscape, it is worth reviewing some policies that have an impact on sinkholes. Specifically, rules regarding water-retention ponds, wetlands, and water storage and recharge will be reviewed.

Water-retention ponds are used to store surface water that flows across the landscape. Prior to the Clean Water Act and the various interpretations and modifications of the act, surface water was diverted in cities like Miami and Jacksonville through storm-water sewer systems directly into surface water bodies like the Intercoastal Waterway. When it was determined that it was not in the best interest of the environment to dispose of storm water in that fashion due to pollution (particularly nutrients), water catchment systems were built in the form of storm-water retention ponds. In the early days, the ponds were built as if a rudimentary geometrist drew them. They were square or rectangular structures that clearly were designed on a drawing board. Now they are designed with sweeping curves in order to better fit the natural landscape. Indeed, they are sometimes designated as amenities in new developments, and homes are built around them to afford waterfront views.

In the early days of water-retention pond building, there was no real attempt to build ponds separate from sinkholes. Although it would seem logical to use natural depressions like sinkholes as water-retention ponds, it is not environmentally sound to do so. As noted earlier in this book, sinkholes typically have a direct connection to underlying aquifer. The porosity and permeability of the zone of raveling enhances water flow when compared to the area surrounding the sinkhole. Thus any water within a sinkhole can readily infiltrate into the aquifer where it may be drawn for human consumption.

Many of the original water-retention ponds were built with direct connections to the aquifer system (figure 7.3). Florida Administrative Code 62-25 provides rules for storm-water management in the state and delegates the oversight of the rules to local governments or water management districts. The rules are complex, but they provide specific conditions under which storm-water retention ponds can be built. Local governments can strengthen the state rules and often add particular details such as landscaping requirements. Within the rules are particular guidelines for the drainage characteristics of ponds. The underlying soils must have a specific porosity and permeability. The rules also allow for the use of wetlands for storm-water storage under specific conditions, though the wetlands must be protected from extremes of water flow and the water must be of a particular quality before entering the natural wetland. Before obtaining a permit for constructing a wetland on a karst environment from a local government or water management agency, the site must be studied and the site plan must take into consideration the unique characteristics of the subsurface.

Site plans are developed with the goal of creating the best management practices (BMP) for the site. Local governments have a great deal of flexibility in approving or denying storm-water plans based on whether or not a site achieves the BMP goal. For example, if a wetland is available for storm-water discharge, but a more suitable site for storm-water storage can be found on the property, it is likely that the local government will require the use of the nonwetland site. Thus storm-water plans are reviewed on a case-by-case basis when applications are put forward for storm-water

Figure 7.3. A typical water-retention pond at a residential complex in Tampa.

management. In addition, local governments may vary in their approach to the interpretation of state rule. For example, rules for some counties encourage the use of wetlands for managing storm water while others do not. Also, local governments can require setbacks away from the edge of stormwater ponds in order to enhance the viewshed of the property. Specific site development rules may also be required, such as the types of buildings that can be built adjacent to the pond or the types of treatment structures that must be built as part of the system.

It must be noted that many ponds are connected to each other through drainage pipes or weirs, which means that the water that enters one particular pond may move great distances through the interconnected drainage systems. Surface water travels great distances not only through the built surface drainage systems but also through the underground karst system. It is important that our society is vigilant about the type of material that enters surface waters.

Sinkholes and Wetlands

Sinkholes are often protected by wetland rules. In Florida, as in the rest of the country, wetlands were drained throughout the nineteenth and early twentieth century in order to encourage development of agricultural and urban land. A variety of government programs assisted landowners in draining wetlands for agricultural conversions. It is believed that of all the wetlands that existed in the United States, approximately half were lost prior to the development of wetland rules (Mitsch and Gosselink 1986), with significant losses in Florida. Wetlands are highly productive ecological areas. They have a tremendous biomass. When wetlands are destroyed, the plants and animals are typically killed or displaced in the process leading to a net reduction in biomass and ecological productivity. Duck hunters saw this very keenly in the early twentieth century. As wetlands were drained, hunters noticed a distinct decline in the game bird population. Hunters' desired birds were creatures that depended upon wetlands for food, nesting sites, and safe havens.

The hunters and their allies pushed for the protection of wetlands in order to preserve the nesting grounds of the game birds. Their allies included a wide variety of conservation groups and local activists interested in preserving land. By the early 1970s, during the major push for widespread environmental legislation in the United States, concern for wetland protection resulted in concrete laws that enforced their preservation. There are dozens of federal, state, and local laws that protect wetlands in Florida.

The Clean Water Act of 1972 and its subsequent amendments are perhaps the most far-reaching of the rules implemented by the U.S. government to protect wetlands. In this act, the government prohibits the discharge of materials into wetlands that would compromise their ability to function as a natural system. The Army Corps of Engineers is responsible for ensuring that this rule is followed, although the Environmental Protection Agency is given authority to reject proposed projects that could impact the natural resources of the site (Votteler and Muir 2002).

Another important development in wetlands protection was the implementation of the so-called Swampbuster Act, which was part of the Food Security Acts of 1985 and 1990. Up until this point, farmers were typically encouraged to drain their wetlands and convert them into cropland. With Swampbuster, farmers were encouraged to convert their wet areas into permanent set-asides in order to preserve them as wetlands. This was a radical shift in policy. Instead of helping to convert wetlands to dry lands, the government now helped farmers convert dry lands to wetlands for long-term conservation—a sweeping shift in policy and an important one as the vast majority of wetlands are located on private land (Votteler and Muir 2002).

Once it was determined that it was useful to protect wetlands, new guidelines needed to be developed that defined exactly what was and was not a wetland. Given that wetlands are real estate, the exact definition of a wetland became a highly charged issue. Indeed, wetland delineation is still a contentious issue that sometimes requires parties to seek legal assistance. The first major wetland delineation guidelines provided by the federal government was the *Corps of Engineers Wetland Delineation Manual* (U.S. Army Corps of Engineers 1987). The manual defined wetlands by different categories, including marine, estuarine, riverine, lacustrine, and palustrine systems. Of particular note are the lacustrine and palustrine wetlands. It is in these wetland categories that most Florida sinkholes are found.

Lacustrine, or lake, wetlands include wetlands of the shore and open water. This type of wetland is common in Florida. Often the shores of the lakes include forest vegetation such as cypress trees or some type of grass or sedge. In Florida, this wetland type is transitional with palustrine wetlands.

The palustrine wetland category includes a variety of different types of wetlands, all of which are classified as having some type of vegetation cover. This category includes wetlands that are commonly called swamps, bogs, and marshes. In Florida, palustrine sinkhole wetlands include vegetated (with grasses or trees) seasonally and permanently flooded depressions. One unique type of palustrine wetland found in Florida is the cypress dome (figure 7.4), a depression forested by cypress trees. The oldest and largest trees are found in the center of the depression, with tree size grading from large in the center to small at the edges. Thus, from a distance, the landscape looks like a dome of trees, although the topographic expression is the exact opposite of the appearance of the elevation of the trees. The palustrine wetland category also includes the vast area of the Everglades in south Florida. The exact type of wetland that exists in any given area is very much dependent upon the hydroperiod of flooding (Keddy 2000). Palustrine wetlands typically are not flooded for the entire year and thus have a distinct dry season.

Due to the extreme variation in moisture conditions, there is often heated debate over exactly where the edge of a wetland is located. During the development process of any land, the state requires an evaluation of the land to determine if any wetlands exist on the property. If they do exist, the

Figure 7.4. A cypress dome in west-central Florida.

landowner is typically restricted from developing in those areas. Because of this, the exact location of the line delineating the wetland areas can greatly impact the choices the landowner has in developing his or her property.

Wetland delineation is done using a combination of field evidence, including the presence of particular plants and hydric soils and the local hydrology. Often, the evidence is not clearly apparent to the landowner, particularly during the dry season. Prior to wetland delineation, homes could be built in areas where palustrine wetlands are found. Indeed, some of these homes were built during dry seasons within sinkholes even though the sites can flood during wet years. Wetland delineation would have prevented the building of homes in these areas.

After the implementation of the wetland rules, it became evident that some of the rules did not make sense. For example, a single palustrine wetland protected in the middle of a vast parking lot does not really function as a viable wetland and does not have a functional value as an ecological entity. The scientific community began to think more holistically about preserving wetlands and their surrounding areas. Several different approaches were taken to alleviate the problems of preserving isolated wetlands in an urban landscape. The first of these approaches was wetland mitigation. Mitigation allows the destruction of a wetland if a wetland of similar or larger size is created to replace the lost area. The mitigation site could be on property or somewhere off property. One of the goals of mitigation was to encourage construction of wetlands in places that would allow for connectivity of ecosystems. Unfortunately, many mitigation projects were not that successful (Votteler and Muir 2002). It is extremely hard to re-create wetlands due to their complex and interactive hydrology, soils, and vegetative communities.

A new approach was developed in the 1990s called wetlands banking. In this approach, developers can pay a fee for the destruction of wetlands in their area, with the fee going for the protection of larger wetland bodies or the construction of large wetland systems. The fee allows land managers to purchase ecologically significant swaths of land that promotes the protection of not only wetlands but also their surrounding upland areas.

The federal government has largely given over the protection of wetlands to state and local governments. In Florida, the water management districts and the Department of Environmental Regulation often manage wetlands rules. Some of the state's wetland rules are stricter than federal guidelines, and it is the responsibility of the above agencies to ensure that landowners follow both federal and state regulations. In addition, local governments can provide more detailed regulations, such as size of buffer zones between wetlands and development areas. Local permitting agencies must ensure that local wetland rules are followed in these circumstances. These rules can be highly controversial and are often challenged by local developers. In recent years, efforts have been made by regulators to include a variety of stakeholders in the development of rules and in the permitting process.

As noted elsewhere in the book, Wilson (2004) mapped sinkholes in Pinellas County that were lost as a result of development between the 1920s and 2000. She found that hundreds of sinkholes were filled and used as sites of development in the last eight decades. If today's wetland rules were in place throughout the twentieth century, many if not most of these sinkholes would have been protected from development. Certainly the loss of sinkhole wetlands in Pinellas County is similar to losses we have seen in other communities, such as Tampa and Orlando. Wetlands rules developed over the last few decades clearly have had a significant impact on the protection of sinkhole wetlands in the state.

Sinkholes and Well Fields

Many communities protect groundwater watersheds where large well fields are located. As noted elsewhere, the state is divided into particular water management districts that have the responsibility of permitting or denying water projects in communities to ensure that the resource or the environment is not degraded and that well projects do not deleteriously impact wetlands or cause other damage to the ecosystems, including sinkhole wetlands.

In the late 1990s and early 2000s, Florida experienced one of the worst droughts on record. Rivers, such as the Withlacoochee, dried to a trickle. Wetlands were parched and lakes disappeared. Water table levels fell to historic lows. Hundreds of brush fires occurred, and many private wells went dry. The government response to the drought was strong. Watering restrictions in communities were imposed and water use was greatly curtailed. During the worst of the drought, it was noticed that the areas set aside for well-field protection experienced significant problems. Many of the wetlands went completely dry and, in some areas, the land sank. The northern Tampa Bay well fields, which supplied a significant amount of water to the Tampa Bay area, experienced considerable damage from the drought, and many people questioned whether or not the well field would be sustainable as the region grew.

At the start of the drought in 1998, a new governmental organization called Tampa Bay Water was formed to help to manage the water in the Tampa Bay area. Prior to the formation of this organization, each of the communities in the bay area was responsible for obtaining their own water. Some communities worked together, but there was no real effort among the urban areas of the region to coordinate water extraction and usage. However, with the formation of Tampa Bay Water, six government organizations worked together: Hillsborough County, Pasco County, Pinellas County, New Port Richey, St. Petersburg, and Tampa.

One of the key goals of Tampa Bay Water was to protect well-field ecosystems. The Southwest Florida Water Management District noted that the pumping of some of the fields significantly damaged the environment, and there was concern over whether the well fields could adequately provide enough water to the growing Tampa Bay region. In response, Tampa Bay Water undertook an ambitious agenda of greatly increasing the amount of water available to the area without developing new well fields. Two of the projects, a desalination plant and a large reservoir, were highly controversial.

The Tampa Bay desalination plant is the largest such plant in the United States. It has a unique design in that it uses cooling water from a neighboring power plant. The water is processed through reverse osmosis. Briny waste material is mixed with more cooling water to dilute the waste before it is discharged into Tampa Bay. The plant has been plagued with problems, largely as a result of filtration problems and membrane clogging, but it now produces 25 million gallons per day of freshwater, which is roughly 10 percent of the area's drinking water.

Another innovative project, the C. W. Bill Young Regional Reservoir, was completed in 2004 in southeast Hillsborough County. The reservoir draws water from the Tampa Bypass Canal, the Hillsborough River, and the Alafia River during high water flow. It can hold 15 billion gallons of water. The reservoir stores excess water that runs off during the wet season for use during the dry season. The project has been the focus of some controversy as there is concern that the reservoir, which is constructed with earthen sides, could fail and inundate the surrounding residents. In addition, there is concern about the change of ecology to the downstream portions of the rivers from which the water will be drawn.

Regardless, these projects, and others implemented by Tampa Bay Water, have the potential to greatly improve water management in the bay area. However, as the region grows, there will be more pressure to develop the groundwater reserves in the region, which could harm sinkhole wetlands and lakes. In addition, several years ago a proposal was put forward to develop a pipeline from the water-rich areas of north Florida to the dryer portions of south Florida. No action has been taken on this proposal, but there will certainly be great debate on this topic if it ever moves forward as a water management option.

Cave Protection

It would be inappropriate to leave this chapter on policy without discussing cave policy in some way. After all, sinkholes often lead to caves and are part of the overall interconnected karst system in Florida. There are hundreds of caves in the state, only one of which is commercial—Florida Caverns State Park. This park is located approximately fifty miles west-northwest of Tallahassee and is a great place to see the type of cave formation that exists in subsurface Florida (the remainder of the state's caves are noncommercial and difficult to access). Caves form under saturated conditions as water flows through conduits in limestone. Cave volume expands with time. If the water table drops to expose the cave to air, secondary mineralization creates speleothems as mineral-rich water percolates through the overlying rock to drip into the cave. With time, large speleothems such as stalactites and stalagmites form.

The caves in the state can be divided into two broad categories—flooded caves and aerated caves—though of course there are some caves that are both flooded and aerated. The flooded caves are saturated by water most times of the year and are accessible only by cave divers. These caves have very few stalagmites and stalactites because these features form under unsaturated conditions as mineral-rich water filters through the cave. Many of these caves can be entered at springheads. Unfortunately, many divers have lost their lives exploring these types of caves, and cave diving remains one of the most dangerous activities that one can undertake in Florida. However, it is important to note that these caves are increasing in volume with time as water dissolves rock in the cave walls.

Aerated caves are scattered throughout Florida but are most common within the ridge lands of the central portion of the state (figure 7.5). Many caves have been mapped and explored in the Brooksville and Ocala Ridges.

Figure 7.5. Florida caves tend to be relatively small with abundant sediment. Photo courtesy of Jason Polk.

Here, many of the caves are indicative of paleokarst formation when sea level was higher. After the sea level dropped, speleothems formed. Currently, the caves are in the process of collapse. In fact, many of the caves in Florida have distinct evidence of roof collapse features.

Unfortunately, a great number of caves have experienced significant vandalism in Florida. It is hard to find a cave that has not been damaged in some way (figure 7.6). Spray-painted epithets, broken speleothems, and litter are common forms of vandalism. Many states have developed laws against cave damage. Florida law (Section 810.13) forbids any kind of vandalism to caves (van Beynen and Townsend 2005; van Beynen et al. 2006). In addition, it is against the law to take speleothems from caves, sell them, or transport them across the state line for sale in another location. Also of importance is the fact that it is illegal to kill, remove, harm, or disturb any naturally occurring cave

Figure 7.6. Many caves in Florida are popular with amateur cavers. Unfortunately, many caves suffer significant damage because of overuse and abuse. This one has been marked with graffiti. Photo courtesy of Jason Polk.

life. This rule is important because many cave creatures are quite rare and deserve special protection.

Before closing this section, it is important to mention the good work done by many of the cave organizations in the state. Several groups, known as grottos, are quite active in Florida. Part of a loosely knit national organization of cavers, they are encouraged through the National Speleological Society and are great sources for local karst information as they often have some of the best cave and sinkhole maps available. In addition, grottos in Florida are involved in volunteer efforts to remove trash from sinkholes, and they sponsor many caving experiences for anyone interested in learning about Florida caves.

Summary

As the population of Florida continues to grow, people's exposure to sinkholes will increase. We are just now coming to grips with the realities of the cost of sinkhole damage to both the public and various sinkhole-related industries. Our insurance rules provide some degree of coverage for property owners, but there is still confusion over what exactly is covered—as well as over the very definition of a sinkhole. Moreover, we have many problems with sinkholes and water. It must be remembered that sinkholes are important wetlands and vital conduits for water. How we manage our water directly impacts sinkhole ecology.

8

Evaluations and Repairs

Over the years I have fielded many questions from the public about sinkholes. Often, a caller or visitor has just found out that their home has in some way been damaged by a sinkhole and they are nearly in a panic. They don't know what to do and they don't know where to begin the process of repairs. They also don't know whether or not to make an insurance claim because they don't know whether making a claim might affect their investment in their home. This chapter reviews the insurance claim process, including what geotechnical firms do when investigating sinkhole claims. It also discusses several approaches to stabilizing property to prevent further damage. Sinkholes have an impact on so many people, and they often do not know where to turn or what the implications are. I hope this chapter will assist them in some way.

Making an Insurance Claim

Imagine the following scenario. One morning you wake up and notice that your bathroom door does not shut right. Also, a crack that was in your ceiling is wider than it was, and some dust from the crack fell overnight onto the floor. You also notice that your terrazzo floor has developed a large crack that extends the length of the house in radial fashion. Upon closer examination, you see that several other smaller cracks parallel the largest one. Could it be just the house settling? After your shower, you leave the house to go to work and notice a slight depression extending from the edge of the house into your yard. Clearly, the ground shifted overnight and something is terribly wrong. Now what?

This scenario is typical of what homeowners could expect to encounter with a sinkhole problem. Some of the expressions of sinkhole damage can be much more subtle—in the form, for example, of some minor cracking in a driveway, wall, floor, or ceiling. Or the damage could be much more substantial: an entire house collapse.

The first thing one should do if one suspects that a sinkhole is damaging a home is to evaluate the situation for safety. Although sinkholes rarely cause loss of life, one must be concerned about the stability of the home. Obvious signs of instability are floors giving way or tilted walls. If that is the case, everyone within the home should get out as soon as possible and call 911. However, most sinkhole events do not require such a rapid response, and most homes can be lived in while evaluations and repairs move forward.

After ascertaining whether or not the home is stable, the homeowner should call his or her insurance carrier. At this time, the insurance company will typically send out someone to do an evaluation of the home to determine if a full sinkhole evaluation is warranted. Sometimes, the damage is so obvious that a full evaluation is not necessary. In these cases, a void space is visible beneath the home and the home has sufficient damage to warrant moving to the repair evaluation. However, in most cases, the insurance companies will require some type of geotechnical evaluation in order to determine the cause of the damage.

As noted elsewhere in the book, there are many causes of structural damage to homes. Buried organic matter that oxidizes with time is a major source of structural instability, as is the subsurface presence of shrink-swell clays. In addition, poor construction practices can hurt homes in the long-term and mimic some of the damage caused by sinkholes. One of the stories that I heard as I talked to people about this book is of a homeowner who had significant structural damage from what he thought was a sinkhole. Upon further investigation, it was found that a broken underground pipe in his lawn watering system washed away the ground supporting one side of the foundation of the home. Thus it is important that insurance companies investigate the exact cause of the property damage in order to comply with the coverage within the policy. Damage to a structure from a sinkhole may be covered, though many other causes of damage are not. Although many wish we had universal coverage for these other vexing problems, Florida insurance providers are not required to cover these other causes of property damage. It is important to note that while some might not be happy with the lack of coverage for these other causes of property damage, Florida is unique in that it is one of the only places in the country that requires catastrophic ground failure coverage and offers sinkhole coverage. This insurance has helped thousands of homeowners recoup losses from sinkhole damage.

It must be understood that about half of all sinkhole claims are denied. The reason for this high number is that, as noted above, there are a number of different reasons why foundations or walls crack or fail. Thus a homeowner should be prepared to hear news that may not be positive. If this is the case, the homeowner has three choices. He or she can fight the evaluation done by the insurance company, repair the house at his or her own expense, or move out of the house and call it a total loss. Let's examine each of the choices.

If the homeowner decides to fight the evaluation of the insurance company, he will likely need the assistance of an attorney who specializes in sinkhole claims. Attorneys with this specialty are quite common and can be readily found by doing an Internet or phonebook search for "sinkhole attorney." Before deciding on a particular attorney, one should do some homework. How many sinkhole cases has the attorney handled and what is the success rate in those cases? And one should have some information on the geotechnical firms they will use when reevaluating your claim. Once the homeowner settles on an attorney, the attorney may try to settle the case without conducting another geotechnical investigation if the geotechnical report is seen as weak. The homeowner may not get the full value of the home, but he may be able to get a settlement at this point. The benefit of settling at an early stage with a company is that less money will go to attorneys and to geotechnical investigations. Even though the settlement might be less than desired, the homeowner, through early settlement, might actually receive more money than if he'd paid to conduct more expensive tests.

If the homeowner finds that the insurance company will not settle, the next phase in challenging the company is to have a geotechnical report done by a different company than that used by the insurance company. Attorneys experienced with sinkhole claims are also experienced in selecting geotechnical firms. The evaluation that firm performs may or may not support the homeowner's case. If it doesn't support the case, the expenses of challenging the outcome of the insurance company's evaluation will have been for naught. However, if the attorney's geotechnical firm does provide an evaluation in support of sinkhole damage, there is a good chance the insurance company will settle once it is presented with this new report. Sometimes insurance companies do not wish to settle even when a geotechnical report is in favor of sinkhole formation. When this happens, the homeowner can decide either to walk away and take the loss or continue to fight through arbitration or other court activity. At this point, the quality of the geotechnical evaluation matters greatly to the individuals reviewing the case. That is why it is critical that the homeowner evaluates the attorney and geotechnical firm that will handle the case prior to starting the process. With a questionable geotechnical report, it is likely the case will be lost. It is better to be told by an attorney that the case is poor at the outset than to tilt at windmills at great expense.

Let's examine our second option. If a homeowner opts to repair his or her home, a home improvement loan will probably be needed if cash is not at hand. Depending on the circumstances, the owner's mortgage lender may assist in providing a loan, or the owner can find a loan somewhere in the financial market. In order to preserve the value of the home, it is important to complete repairs to the foundation and structure. Any patch job will likely be seen by future home inspectors and, thus, the value of the home will decrease over the long haul. Also, it is appropriate to have a geotechnical engineer evaluate the home to determine if the problem that caused the damage is going to occur again or if it was a one-time event. If the problem is likely to occur in the future, it may be harder to conduct appropriate repairs unless the subsurface problem can be removed. The geotechnical engineer can assist in evaluating options if the problem is recurring.

The final option, walking away from the home as a total loss, is one that should be exercised if the financial costs of repairs are beyond one's means. Dozens of companies in Florida purchase damaged homes. Many of them will write a check within days of striking a deal, enabling homeowners to get out of the situation quickly and without too many headaches. Unfortunately, in this case, the homeowner often loses the accrued value of the home. However, these companies often provide a fair market value for a damaged property. One should get several estimates before deciding on a company. As noted in a previous chapter, these companies are often able to repair the homes and put them back on the market in order to make a profit.

No matter what approach one takes to dealing with a damaged home, the process is painful. There is no denying that the process of putting forward claims and conducting house repairs is frustrating and time consuming, but this is the price some of us pay for living in a beautiful state that happens to be underlain by karst voids and other subsurface features.

The Geotechnical Evaluation Process

Geotechnical evaluations are regularly conducted in Florida as part of sinkhole insurance claims evaluations. There are dozens of companies in Florida that do this type of work, and they can be found online by searching for "geologists," "engineers, environmental," or "engineers, geotechnical." There are four main evidence types in the geotechnical evaluation process: evidence of previous sinkhole activity in the area, evidence in the field, evidence from borings, and evidence from geophysical investigations. An analysis of each research strand allows the engineer or geologist to determine whether or not a sinkhole is present on the property. Each of these types of research is discussed below.

Previous Evidence

Prior to completing any field investigations, the geotechnical firm will conduct a detailed investigation into the potential for sinkhole formation at the site and on the geology of the region. This will be done through map and data analysis. Topographic maps are evaluated for the presence of topographic sinkholes in the area. In addition, the company will review where reported sinkholes occurred in the past. This can be done in three ways. First, county engineering departments typically keep a database of sinkhole foundation problems. These lists are public documents and can be reviewed by firms to locate where other sinkholes occurred in the area. In addition, the firm will likely review the state sinkhole database maintained by the Florida Geological Survey. Second, many geotechnical firms maintain their own database of sinkhole activity. Because such lists are privately owned and are not available for public review, there is likely a great deal of variability from one list to another. These lists may include sinkholes that were officially reported to county governments and other sinkholes that may have formed in the area that were not part of an official report. Third, some insurance companies maintain a database of sinkhole locations. Again, these lists are privately owned and inaccessible to the general public.

The locations of sinkholes in the area in question will give the investigator a sense of the likelihood of sinkhole activity in the area. If there have been numerous sinkhole reports in the region, it increases the likelihood that a sinkhole will be found. However, if other investigations found nonsinkhole causes for structural damage in the area, the investigator will carefully examine other possible causes for the damage while keeping an open mind regarding sinkhole damage. In other words, the locations of various forms of damage provide a sense of the type of geologic activity that typically occurs in the area. While investigators remain open to all possibilities, the variation in the causes of foundation failure provides some guidance.

If the company conducted investigations in the area in the past, previous reports will be reviewed to get a detailed sense of the local geology. In addition, geological reports of the region, if any, will be reviewed. The investigator will also collect aerial photos of the site and any other type of maps that may be available, and that can assist in interpreting the geology and karst geomorphology of the area.

Field Analysis

When an evaluator visits the site, she will look for field evidence of sinkhole activity. The activity might be quite obvious and in the form of a distinct depression on the property. Indeed, the depression may intersect with the foundation of the house. The evaluator will examine the depression in order to assess if it is caused by karst activity or by some other factor. Typically, the evaluator will record the dimensions of the depression and record the feature's roundness and the steepness of the sides. In addition, the evaluator will look for cracking in the ground or other features on the property, such as driveways. Sometimes when a sinkhole forms, cracks parallel to the trend of the circular depression form. These are seen as a series of radial cracks around the main depression and may indicate wider settlement problems.

Also, the inspector will examine the property damage. Sometimes the damage is in the form of cracked foundations or walls, while other times the damage may be more substantial. Regardless, the inspector will examine the damage for clues as to whether or not the property received damage from a sinkhole. Particularly useful to the inspector are patterns of cracks that may be found in the foundation. Radial cracking is most indicative (although not definitive) of sinkhole formation. The inspector will check doors and windows to evaluate structural damage to walls.

Boring Analysis

An evaluation will typically consist of two different types of boring analyses: hand boring and mechanical boring. Hand boring is done with a hand-held probe, a hollow tube that allows the evaluator to collect near-surface soil samples. Depending on the type of probe used, the evaluator can typically get information on the upper 1–4 meters of soil. The reason for the hand probe is that the soils above a sinkhole are often very loose due to a recent disturbance. Thus, when probing is conducted, the evaluator is trying to find the presence of loose sediment on the property that may be indicative of a zone of raveling. As noted elsewhere in the book, the raveling zone is an area in the sediment directly above a subsurface void that is impacted by filtration of sediment into the void. The zone is often mixed and is much looser than the surrounding sediment. In addition, small voids may be found within the zone. Because of these characteristics, hand probing often allows the evaluator to assess where a sinkhole problem may be located on a property. The evaluator will record the type and consistency of sediment present at different boring sites. The boring sites will then be mapped, and the variations of the boring analysis will assist in determining what type of subsurface problem exists at the site.

Mechanical boring is done typically after all other investigations have been completed in order to confirm the presence of subsurface sinkhole activity that might have been suggested by the hand boring. Geophysical investigations, discussed below, provide good images of the exact site where sinkhole activity probably took place. Mechanical boring confirms the presence of the sinkhole by probing directly into the raveling zone or void.

Boring is done using some type of mechanical drill rig—typically a hydraulic system that forces a pipe into the ground. The operator records the way the pipe enters the ground and the sediment present in cores collected in the process. The reason it is important to record the way the pipe enters the ground is that raveling zones or subsurface void space allows the pipe to rapidly move through the subsurface compared with other subsurface material. Thus if a drill encounters a void, the pipe will suddenly drop through the void without notice. Such a drop is an indicator of the presence of a sinkhole. If the pipe encounters loose sand, which is associated with a raveling zone, the pipe will also move rapidly through the area. Skilled drill operators are able to discern these telltale features based on how the pipe reacts to the subsurface.

The operator will also record sediment collected in the process by depth. This step is crucial because certain sedimentary deposits can cause ground instability and lead to problems that mimic sinkhole formation. As noted elsewhere, these deposits include buried organic matter and shrink-swell clays. If these unstable materials are present, without the presence of a void or raveling zone, it is likely that a sinkhole did not cause the foundation problems of the home.

Geophysical Investigations

Perhaps the most important part of the overall geotechnical assessment is the geophysical investigation. Geophysical tools are used to get a three-dimensional picture of the subsurface geology in order to determine whether or not voids or raveling zones are present. A number of different types of geophysical approaches can be taken in conducting an investigation, but the most common approach is ground-penetrating radar, a technique that works by pulsing radar waves into the ground and measuring the subsequent reflection of the radar waves by features in the subsurface. Most GPR systems work by dragging or carrying a transmitter across the landscape in grid fashion (many GPR devices are quite small and are easy to maneuver across a property).

As the GPR device is moved across the property, the radar waves are pulsed into the ground and reflected back into a receiver. The receiver then collects and stores the data and processes it for visualization. The device also has a display unit that can show real-time results as a three-dimensional view of the subsurface features. The displays are difficult for the layperson to interpret, as they typically look like a series of irregular lines, but with practice, an operator can find distinct subsurface geologic layers, changes in moisture conditions in the soil, and irregular karst features such as raveling zones and voids.

As an operator works, he or she looks for the raveling zones and voids in the display unit. These are the signs of sinkhole activity that allow the geotechnical evaluation process to move ahead with mechanical boring and interpretation of sinkhole activity as the cause for structural damage. The exact location of the raveling zones or void space is important in order to locate where the mechanical boring will be done on the property. Mechanical boring is quite expensive and GPR helps to reduce costs by providing specific locations where the boring should take place to verify the presence of voids or raveling zones.

The use of GPR and mechanical boring confirms or rejects the claim of sinkhole activity. As per state definition, in order for damage to a home to be covered by standard sinkhole coverage, the sinkhole must be found on the property and found associated with the damage. Thus if GPR and associated mechanical boring do not find a raveling zone or a sinkhole on the site, the insurance company will likely reject the sinkhole damage claim.

If there is a positive assessment for sinkhole damage, GPR and the mechanical boring become quite important in the assessment of site stabilization. As we will see, a number of approaches can be taken to stabilize a property after sinkhole damage. The nature of the subsurface anomaly is important when it comes to how to manage the stabilization and repair effort.

The final report produced by the geotechnical firm will include all of the information produced in its investigations. The background information obtained through archival investigations of maps and previous reports in the area help to place the damage in a geologic and regional context. The field investigations allow the investigator to evaluate the evidence that is present in the field in the form of topographic anomalies, damage, and near-surface sedimentary irregularities. The mechanical boring allows investigators to map exactly where the anomaly exists and confirms geophysical investigations that provide a detailed three-dimensional visualization of subsurface conditions. Combined, this evidence allows investigators to (1) evaluate whether or not the damage is caused by sinkhole activity or if it is caused by other factors and (2) provide suggestions for stabilization and repair if a sinkhole caused the damage.

Interpreting the Geotechnical Report

Over the years, I have received a number of calls from people who have had some type of geotechnical investigation of their property and cannot interpret the results. They have read the summary of the report that supports or rejects the notion that a sinkhole caused the damage, but they do not have a clue as to how to understand the rest of the report. While I would love it if more people were better educated on Florida geology, I must admit that the technical nature of the reports is confusing to the average person. The reports include abundant detail about subsurface sediments and drilling logs. In addition, the information on the geophysical investigations is particularly difficult to understand—especially the figures that show the results of GPR investigations. While those in the insurance industry who work with this information on a day-to-day basis have no trouble reading and interpreting the reports, the information contained in them is largely incomprehensible to the general public.

Unfortunately, due to the technical nature of the reports, the average person is unable to read each word and understand exactly what the entire report means, no matter how clearly written. Thus it is crucial that these reports have a clear executive summary that reviews all of the work completed and the results of that work. Many reports do not live up to this goal. The executive summaries are often brief and discuss whether or not a sinkhole is present, not the specifics of the results. For most homeowners, a brief summary of things they should look for in the report that will help them understand the geotechnical details and the meaning of the results would be helpful. Below is a list of results common in geotechnical evaluations of sinkholes with descriptions of how to interpret the results. The list is not inclusive, as there are no standards for preparing geotechnical reports, but it does contain the most common elements found in such reports.

Location of other sinkholes in the area. The geotechnical firm should evaluate whether or not sinkholes are common in your area. They should also examine whether other causes of foundation failure have been reported. If multiple reports of foundation failure due to sinkhole activity have been reported for your vicinity, there is a likelihood that your damage is also sinkhole related, and this information helped the geotechnical firm evaluate the site. If other causes of foundation failure, such as buried peat, are common in the area, this will be reported, and the likelihood that your foundation failure is sinkhole related is not high. Geotechnical firms will still complete a full sinkhole evaluation in these circumstances, but the information helps them understand the local geology. In the unlikely event that there are no foundation failures in the area, a full sinkhole evaluation will move forward, but there is a possibility that the foundation failure is caused by poor construction practices or other causes. The report will typically list addresses or specific areas where foundation failures have been reported in the past. Sometimes the location of these failures will be shown on a map or aerial photo.

Geology of the area. Typically, the reports contain a brief description of the geology of the area where the damage was reported. This information is limited to the surficial geology associated with the active zone where karst processes influence the surface stability. The geologic descriptions will include a description of surface sediments and bedrock. As noted in chapter 2, unconsolidated marine sands underlie much of Florida. Beneath these sands is a layer of clay that separates the sands from limestone bedrock, and karstic voids form within the limestone. When sinkholes form, the overlying rock collapses into the voids to cause a depression on the surface, or sands slowly filter into the void through a raveling process to cause a depression on the surface. There are subtle but important local variations to this simple geologic schematic just outlined. The report should include the local geologic setting that reflects the distinctive local geologic character of the region. Typically, the variations include particular layers of clay or organic matter within the unconsolidated material overlying the bedrock. These layers are often in the form of small lenses within the broader, more regional sedimentary layers. The lenses indicate a small environment of formation such as a temporary lagoon or a lake. Lenses of organic matter or shrink-swell clays can be especially problematic to homeowners because these lenses can cause instability at the surface. The geotechnical report should include a description of the size and composition of known geological lenses in the area.

Field evidence at the site. The report should include an assessment of field evidence found at the site. The investigator will look for active depressions and measure their size and shape and map their location. In addition, the report should contain a detailed description of the damage seen at the site. Here it is important that the homeowner inform the geotechnical firm if something is not accurate or if something is not included. The homeowner should be present when the geotechnical investigation takes place so that he or she can point out to the geotechnician the exact damage and any depressions that might be present on the property. In some cases, when field evidence points to a genetic cause of the damage, the report will include a description of the probable causes of the damage. However, most situations are not that clear and thus the field evidence is reported without comment as to cause.

Hand probes. Typically, the technician completing the assessment will conduct a series of hand probes at the site. The results of these hand probes are included in the report and are listed by number and are mapped. The probes are described by the sediment type encountered and the consistency of the soil. Soil consistency is defined as how the soil holds together. Soil consistency can vary from loose to hard. What the technician is looking for in conducting soil probes is whether or not the soil is particularly loose in certain areas and whether or not void space can be encountered through probing. When the sinkhole forms beneath the surface, the raveling process disrupts overlying sediment, allowing it to become loose compared to the surrounding sediment. Soil probing attempts to find such loose areas. When they are encountered, they are mapped and described. The soil probe descriptions also include a statement of the sediment type encountered. There are three main sediment sizes in soils (in descending size): sand, silt, and clay. Sands are noncohesive, have a high porosity (room to store water) and permeability (ability to transmit water), and are the most common sediment at the surface in Florida. Silt is fine, like flour, is not very common in Florida, and has a moderate porosity and permeability. Clay is extremely fine, is common in lenses within sedimentary bodies in Florida, and is found at the interface between the limestone bedrock and overlying sedimentary units. Clays have a high porosity, but extremely low permeability, making them unable to transmit water very well. Some clays in Florida also have shrink-swell properties that enable them to expand to many times their dry volume when wet.

Geophysical investigations. Perhaps the one area of sinkhole technical reports that is most foreign to the general public is the geophysical analysis. Most of the time, one technique, ground-penetrating radar, is used. As noted elsewhere, GPR is a technique that pulses radar waves into the ground. The images produced as a result of data processing and visualization are utilized to interpret the subsurface geology. To the untrained eye, GPR images look like a bunch of zebra-striped lines on a page. However, a skilled analyst can discern sedimentary variations, groundwater variations, and the presence or absence of sinkholes. A readout of a location scanned for sinkholes shows a cross-section of earth measured along a transect using GPR. An absence of sinkhole activity displays as neatly horizontal lines that show little variation except in line thickness. The striations in the readout might show very slight disparities associated with moisture conditions, but otherwise the GPR transect will appear quite uniform. By contrast, figure 8.1 shows a distinctly uneven GPR readout. Instead of a pattern of straight, neat striations, the readout shows definite disruptions in the horizontal lines. These disruptions are geographically distinct along the transect and occur within narrow zones. In addition, where the horizontal lines are broken, the breaks have a vertical directionality. Such breaks are typical indicators of

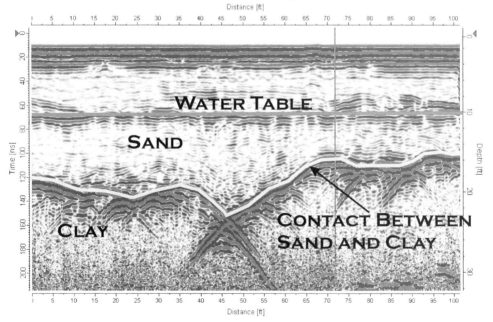

Figure 8.1. GPR readout showing the location of a sinkhole. Image courtesy of GeoView, Inc.

raveling within the subsurface and are clear indicators of sinkhole development. A GPR report should note the location where transects were taken and provide representative GPR images that show the nature of the subsurface. Raveling zones should be noted and mapped.

Mechanical probing. Once GPR investigations are complete, mechanical probing is typically done to confirm the presence of sinkhole activity. Mechanical boring is quite expensive, so the target sites for probing are based on field evidence obtained from hand probing, GPR investigations, or visual field evidence. Boring may also be done within the home in order to assess the specific nature of the material directly under the area of damage. Mechanical boring is done with hydraulic equipment that collects cores of the material in the subsurface. In addition, the technician will note how the probe penetrates the subsurface. If the probe drops some distance without encountering any resistance, the probe has clearly encountered a subsurface void. Technicians must be careful in these circumstances, as some operators have induced sinkholes through the probing process. As with hand probing, the technician will record the sediment type present at the location. The report should include the location where the probes were done, any voids

Figure 8.2. GPR operating where a sinkhole is present. Image courtesy of GeoView, Inc.

encountered in the process, and a description of the sediment type present in the subsurface.

Interpretation of the data. At the end of the report, the geotechnical firm will provide some type of interpretation of the data they collected, including the archival information on previous sinkholes and the various field investigations. In this section, the firm will provide an assessment regarding the presence of a sinkhole at the site. The firm's conclusion is typically based on multiple sources of data. For example, the conclusion should not be based on archival information alone but should also include other positive evidence, such as GPR results or mechanical boring data. Without multiple threads of proof of sinkhole activity, it is difficult to solidly opine as to the presence or absence of sinkholes at the site. When the evidence is scant or absent, firms have a responsibility to indicate that the property damage was not caused by sinkhole activity. Likewise, when there is evidence of sinkhole activity, it is the responsibility of consulting firms to indicate that karst activity likely caused the damage. When sinkhole activity is not found at the property, the report should indicate if any other geologic activity caused the damage. For example, if decomposing organic matter is found in the subsurface, this should help the property owner with decisions as to how to stabilize the subsurface. Unfortunately, as noted elsewhere, there is no standard approach used in interpreting geotechnical reports or in conducting geotechnical investigations and the interpretation of data can vary from firm to firm. Thus when a property owner is not sure about the quality of the investigation or the interpretation of the results, it would be appropriate to ask another firm to interpret the report. There is a cost to this interpretation, but another firm can confirm the interpretation of the organization that completed the report or provide some suggestions as to the quality of the investigation. In some cases, property owners have opted to get a second opinion and pay for a new investigation when they have questioned the quality of the results.

Stabilizing the Ground and Saving the Structure

I recently had the opportunity to talk to a woman who owned a condominium in Tampa that received considerable damage due to a sinkhole. One morning she woke up and found that a large hole had opened in the ground outside of her building and large cracks had formed in her floor and on the walls of her unit. Other people in the building and other nearby condominium buildings were similarly affected. The first thing that concerned the residents was whether the entire property was going to collapse. They also were worried about their possessions, and, in general, they did not know what they were going to do. As it turned out, the condominium insurer moved the individuals and their property out of their units into apartments temporarily until the ground was stabilized and the units repaired. The woman I spoke with was out of her unit for six months while repairs were completed. She was very pleased with how everything was handled, even though she didn't enjoy having to move. But what happened while she was away? How is the ground stabilized and how are property repairs conducted?

Stabilizing the ground

Typically, the first concern after a catastrophic sinkhole, such as the one described above, has formed is securing the property to prevent further collapse and associated property damage. Geotechnical investigations move into a triage operation that attempts to save the existing structure while conducting investigations into the nature of the subsurface void. If possible, some form of geotechnical investigation will take place if it is determined that the ground is stable enough to allow such an investigation. GPR analysis or mechanical boring might be done in order to find out the form and structure of the subsurface void. These investigations should determine whether there is more void space beneath the collapse. If such voids exist, the investigations will also attempt to determine the volume of the voids and the extent of the sinkhole activity on the property. The characterization of the subsurface voids is important in order to assess ground stabilization options. For example, if it is determined that the sinkhole collapse essentially filled whatever void there is underground, there is no need to conduct any particular ground stabilization process because the collapse event essentially ended any risk to the property. In these circumstances, the depression is graded with fill material and the property is repaired. Often, a clay cap is put on the void to divert drainage from the site of the collapse. In addition, the property may be regraded to prevent water from draining onto the site of the collapse (Gordon 1987). However, in many instances, void spaces that pose a continued risk to the property are found in the subsurface. When they are, the best mitigation option is to fill the voids with grout.

Grouting is done not only to fill the void space but also to provide strength to the surrounding bedrock to prevent future instability. In addition, grouting diverts groundwater flow away from the active karst zone, reducing the potential for void formation (Ryan 1984). Grout composition can vary. Typically, there are two considerations associated with the choice of grout composition: volume and strength (Ryan 1984). Because sinkhole voids often require huge volumes of grout, it is important to consider that the cost must be kept low and that the strength of the grout is not particularly significant because it is mainly used as a void fill. In other words, the grout usually does not have to meet any engineering specifications. The grout typically consists of some ratio of filler and cement. The filler can be clean waste products from the mineral extraction industry or other forms of sediment. The cement provides some degree of strength to the fill, although it is not necessary to exceed the strength of the surrounding material. In some special circumstances high cement content is required. In other circumstances, low strength is all that is needed due to the weakness of the surrounding soils (Ryan 1984).

Compaction grouting is a technique that is done to enhance the strength of the surrounding material (Henry 1989). This process is different from basic grouting in that the grouting process creates a pressure underground that makes the surrounding material denser, making the ground more stable. The process can also lift soils or structures due to the pressure exerted in the pumping process.

In cases where collapse is significant, a combined grouting and filling project must be undertaken. Grouting of what is called the sinkhole plug must be done to stabilize the subsurface. A cement pad is sometimes put into place on top of the plug at the base of the collapse to further stabilize the ground surface. On top of the cement pad, fill is then dumped to fill the void. The grout filler and the fill used to grade the depression must be clean as the material will certainly be in contact with groundwater or water percolating to the aquifer.

The timing of the remedial action can be important in saving property. In many instances, geotechnical firms are contacted for evaluation while a sinkhole is still forming. In these instances, immediate action should be taken to stabilize the ground surface. Garlanger (1984) provides examples of how stabilization through grouting and, in one instance, wall stabilization were done very quickly after the initial inspection. Indeed, in one instance, subsurface investigations took place through mechanical boring while grouting was under way.

A great deal of research has been conducted on the stabilization of road surfaces in karst terrains (Villard et al. 2000). Because of their drainage characteristics and the loads that they carry, roads are often sites of sinkhole formation. After a sinkhole forms on a roadway, some type of grouting is used to stabilize the subsurface. However, due to the special characteristics and engineering needs of roads, particular attention must be paid to long-term stabilization of the roadbed. Bonaparte and Berg (1987) detail particular geosynthetics that can be used to stabilize a road surface and support the road should collapse occur again. A study by Louw et al. (1984) suggests the use of reinforced roadway material and associated supports in some particular circumstances. In some cases, attention is given to designing roadways in karst terrain to avoid problems (Moore 1984). Monitoring and assessment of exactly where depressions are likely to occur on roadways also receive some attention (Foshee and Bixler 1994). The problem of sinkholes on paved surfaces can be vexing when sinkholes occur on airport runways (Belesky et al. 1987).

Consideration is also given to protecting large structures in sinkhole-prone areas. For example, Bergado et al. (1984) studied problems with karst landforms in the building of a large dam in Thailand. As Florida's continued growth spurs ever-increasing large construction projects—structures such as the Lee Roy Selmon Expressway, large skyscrapers, water reservoirs, and gypsum stacks—more and better geotechnical investigations will need to be completed to prevent problems.

Building on and in Karst

Building on karst poses a challenge to even the most skilled builders and engineers. It is nearly impossible to plan for every problem that can occur. Indeed, karst activity is unpredictable and can take place almost anywhere in limestone regions. Voids in the limestone can be over 100 feet in width and height and miles long. Thus building on such voids is difficult, particularly as we do not know where they are located. However, several steps that can be taken to evaluate building sites prior to planning any type of development. Tomachev and Leonento (2011) use the term "engineering karstology" to describe the engineering issues associated with construction on karst terrain. A number of engineering techniques can stabilize the ground in areas where, based on these evaluations, there is a likelihood of karst activity. In 1984, Sowers suggested a five-step process for evaluating and preparing property sites for development in karst areas:

1. Optimize the location on the site
2. Correct or mitigate defects that are present
3. Use shallow foundations, modified to overcome the defects
4. Use deep foundations modified to overcome the defects
5. Minimize future activation of defects (Sowers 1984, 375)

Prior to undertaking a major project in karst terrain, it is possible to investigate the nature of the subsurface and any particular flaws that may cause structural damage or other property challenges in the future. This is the optimization process outlined above. In this step, the goal is to find all the karst features possible using all available techniques to help site the structure in a safe location. The techniques used to find karst features may include geophysical approaches, remote sensing, and field observation. Once the karst features have been found, the building footprint can be placed to avoid the more significant karst features. Garlanger (1984) recommends completing a detailed bedrock map for building sites in order to ascertain where subsurface topographic depressions are located. Often, these depressions are filled with Miocene or Pleistocene sediments. These depressions can impact the stability of the subsurface if they are in areas where the piping of sediment to subsurface voids can occur. The bedrock topography also assists in the planning of corrective measures that can be taken to design foundations appropriately.

The correction and mitigation step is completed when it is impossible to avoid karst features in the building process. In these instances, it may be appropriate to employ one of a number of ground-stabilizing approaches, including grouting, installation of subsurface stabilization structures, and filling of depressions. In some instances, sinkhole collapse may be initiated in order to avoid future problems.

Building on top of karst features has many challenges, as Sowers (1984) describes. When building using shallow foundations, one can create a bridge over a potential void, one can use engineered subsurface stabilization techniques, or one can design the foundation for extreme conditions by creating a super-strong and rigid structure. Each of these approaches requires continued monitoring because any structure, no matter how well engineered, has the potential to fail. The use of deep foundations in the form of piles or piers is often done to support large structures. This approach is complicated because there are many different types of deep foundations that are used in Florida to construct large projects. However, it is important to note that while most piers work quite well, care must be used in selecting the depth at which the pier will rest. The pier must sit on solid bedrock that is strong. Care must be taken to not place the pier on weak rock or rock that has significant cracking or karst porosity. In some cases, piers can collapse as a result of rock collapse. Detailed pier and pile studies should be completed prior to construction (Raghu 1987).

Site analysis and construction design requires particular attention when building large facilities on karst systems. Almaleh et al. (1993a, 1993b) describe engineering challenges faced when expanding the Crystal River nuclear facility on top of karst terrain. As one can imagine, the design at a facility like this is important to not only the structural integrity of the structures but also to the broader public health of the people of Florida.

Day (2004) summarizes a variety of problems that were encountered in building a deep tunnel drainage and sewer system in Milwaukee, Wisconsin. In his work, he demonstrates that the designers of the project did not appropriately account for the karstic variations in subsurface bedrock prior to designing the project. Such problems accounted for part of the over-budget expense of $50 million. Similar problems can be encountered in Florida when insufficient subsurface investigations are completed. In the Milwaukee case, 45 percent of the tunnel surface required some type of lining or grouting.

In addition to appropriately designing a project, it is important to grade the site appropriately. It is not desirable to have water draining to karst voids in the subsurface in the property as drainage can enhance void formation and trigger sinkhole formation. Thus grading to divert water away from buildings and karst features is a crucial step in ensuring a safe situation for the building. The evaluation of property prior to its development is important and can prevent serious problems in the future. Indeed, Kuhns (Kuhns et al. 1987) conducted extensive investigations prior to the building of a multi-million-dollar office complex in Maitland, Florida, and filled voids with grout and completed a detailed ground-stabilization plan prior to construction.

In some regions of South Africa, where sinkholes are common, major urban development applications must be evaluated by the South African Geological Survey prior to approval (Roux 1987). The evaluation conducted by the survey in the karst-prone regions requires field investigations followed by a detailed analysis by the developer of the subsurface risks through a program of drilling, mapping, and geophysical investigations. The evaluation provides a mechanism for determining the risk to development in three categories, with risk levels of I and II being acceptable and risk level III being unacceptable. We have not developed such an approach in Florida, although there are certainly differences in sinkhole risk from place to place. The desire for free enterprise and the lack of desire for government intervention in real estate development has negated the development of any enforceable sinkhole risk map. However, as more and more Floridians become impacted by sinkholes, I would not be surprised if at some point there were limits to development in some areas based upon sinkhole risk.

In contrast to Florida, Great Britain has made significant strides in the understanding of the distribution of subsurface voids. Historically, there have been a number of tragedies associated with collapse of subsurface voids. Britain has a number of different types of subsurface voids, including old mine conduits and sinkholes that can collapse to cause property damage and loss of life. In recent decades, the United Kingdom has worked to map the location of both naturally occurring subsurface voids and mine conduits in order to better understand their potential risk to the public (Edmonds et al. 1987).

This mapping has focused on regional risk. Although more than 10,000 natural cavities have been mapped, they are not distributed evenly across the county. Thus, in order to better evaluate the risk, the country was divided into three distinct risk zones: Zone 1 is an area where there is evidence of subsurface voids and cavities; Zone 2 is an area where the geology is conducive to the presence of voids but no voids have been mapped in the area; and Zone 3 is an area where there are no natural cavities present and where they are unlikely to occur. The zone approach to mapping these features allows planners to make appropriate land-use choices and to require particular investigations prior to any land development. In the United Kingdom, this effort took place in the 1980s and provided a great deal of information to land managers.

Unfortunately, we do not have such a map for Florida. While the state's karst geology is vastly different from that of the United Kingdom, such a map could be constructed. Several specific challenges would make creation of such a map difficult but not impossible. The karst in Florida is much more vuggy, or porous, than the karst in Great Britain. Many Florida geologists compare the subsurface void distribution to being something like Swiss cheese. However, as time goes on, we are becoming more aware of the locations of where large voids are present in the subsurface. Yet we are not maintaining a database of this information. As in Britain, there are distinct geographical variations to the distribution of void space in the subsurface of

Florida, and land-use managers would benefit greatly from this information. Perhaps in the future, we will have maps that allow developers and land-use managers to make better decisions about land use in the state. It's just that so many natural and unnatural processes can trigger sinkhole formation. If we know the location of subsurface voids, we can be vigilant about monitoring the land for potential damage. For example, there is evidence of heavy rainfall from tropical storms influencing sinkhole formation (Hyatt and Jacobs 1996). If we know where the risk is, we can be prepared for possible damage during extreme events.

Sinkholes that Aren't Sinkholes

It is worth noting that many sinkholes reported in the media are not sinkholes but are in fact depressions caused by the collapse of underground pipes, such as water lines. When these collapses occur, they often make the news because many of the conduits placed in our urban environment are associated with roadways. Thus, when a pipe collapses, the associated depression typically forms in a roadway, significantly disrupts traffic, and garners eager media attention. Over the years, I have received calls from associates telling me about a sinkhole discussed on the six o'clock news. Most of these turn out to be collapsed pipes, not karstic sinkholes. However, in Florida, and many other areas of the world, the lexicon has changed to the point that the public calls these features sinkholes. These features can be small or large. The larger ones are often enhanced due to flowing water or sewage that washes away sediment to create a large depression.

The term *sinkhole* for these types of pipe collapses has made it into the academic and professional literature. For example, the *Engineering News Record* (1996) reported a story in 1996 about a "sewer sinkhole" that opened on the streets of Brooklyn, New York. A sanitary sewer collection line collapsed 15 feet below the surface of the earth. Although no one was injured in the accident, there was quite a disruption of a busy throughway. Although these types of phenomena are not strictly sinkholes, it is clear that there must be some name for them.

While these are not true karstic sinkholes, it seems prudent to take a cue from the general public and provide a new category of sinkholes for these features. They certainly require investigations and infillings similar to karstic sinkholes that occur in our urban landscape. Also, they don't look all that different from a karstic sinkhole that may form in any given area, although simple observation would definitely enable one to discern between the karstic sinkhole and the one that forms from the collapse of a pipe.

So what should we call these modern-day sinkholes formed from conduit collapse? We could call them "conduit collapse sinkholes," but that is not a good term because the term "conduit" is used to describe a naturally formed karst tube underground. We could call them "pipe collapse sinkholes," but that is not a good term either because "pipe" describes a vertical karst conduit or a raveling zone. Therefore, I suggest calling them "pipeline collapse sinkholes" to indicate the special nature of their genesis. The term could be modified to define the particular type of pipeline that caused the damage, such as a "water pipeline collapse sinkhole," or a "sewer pipeline collapse sinkhole." Each type presents its own unique challenges to those who must repair the situation. The term would clearly differentiate between naturally formed karst sinkholes from those depressions that form as a result of a collapsed pipeline. While it is unlikely that the distinction would mean a great deal to the general public, it would matter to those of us who pay particular interest to the topic. In fact, as I have conducted my research, I have found that some in the engineering field do not differentiate between the types. Their concern is solely repair and ground stabilization. However, when the term sinkhole is used interchangeably in the literature to indicate both karstic sinkholes and those that form as a result of pipe collapse, confusion develops. Thus I urge the use of clear terminology when discussing sinkholes, as their genesis is so very important to their overall interpretation.

Conclusion

Thomas Sputo, in an article he published in the *Journal of Performance of Constructed Facilities* in 1993, provides a fascinating glimpse into what happens to a structure, specifically Chiefland High School in Chiefland, Florida, when it is subject to sinkhole damage. The damage occurred during a rainy weekend in March 1991. On the Friday leading into the weekend, someone in the school felt the building "shudder." Approximately 9 inches of rain fell in the area, and water that had collected in a parking lot drained from this lot beneath the building through the subgrade and into the karst system. Upon inspection, it was found that several parts of the school had been damaged. Walls had separated from floors and structural support piers had settled several centimeters. Geotechnical investigations revealed that a sinkhole had formed beneath the damaged area and that the entire subsurface had the potential for doing greater damage to the school. The author of the report believes that better grading of the site might have prevented or delayed the formation of the sinkhole.

This example illustrates the many problems that can occur in karst settings. Suddenly, the building had significant damage and the school administrators needed to react quickly to determine if it was safe and what they should do to repair the structure. Given what was found in the geotechnical investigations, one questions whether or not the building should have been built on the site in the first place. Certainly, those who built the school in the 1930s could not have anticipated the sinkhole problems, but we now have amazing tools that allow us to have a glimpse into the subsurface. Many problems are currently being prevented by the appropriate use of the geotechnical expertise we have in the state of Florida. With time, I hope that geotechnical engineering will assist in designing better development plans and land-use regulations that can help reduce future property damage.

In 1997, the Florida Geological Survey published Open File Report 72, titled *Geologic and Geotechnical Assessment for the Evaluation of Sinkhole Claims*. The report provides an excellent review of the process of sinkhole evaluations in 1997. One of the report's recommendations is that a sinkhole research resource (SRR) be created. The SRR is really an idea that recommends filling the research void since the Florida Sinkhole Research Institute became largely inactive. The report highlights several areas for sinkhole research that would benefit the state: (1) identification and explanation of sinkhole activity; (2) prediction of future sinkhole activity; (3) centralization and coordination of technical data; and (4) dissemination of information to the general public (FGS 1997, VI-5–VI-6). The report goes on to state that the Florida Geological Survey, the University of Florida, and the University of South Florida have thought through proposals for developing an SRR. To my knowledge, such a group has not formed, although karst and sinkhole research certainly goes on at all institutions. Yet the state would greatly benefit from a coordinated effort to manage sinkhole and karst science and policy research across the whole state. From my perspective, it matters not so much where the work is being done as that every effort is made to enhance communication and interaction among researchers on the topic of karst.

9

Conclusions

I hope I have demonstrated the importance of understanding Florida sinkholes from a science and policy perspective. The landscape of Florida is truly unique in the world. Nowhere else is such a vast area affected by sinkholes and other karst processes. Many Floridians are exposed to sinkholes as the cause of personal or community property damage or by working on sinkhole-related issues in the geotechnical or insurance industries. An ongoing and expanding pool of general information, scholarly literature, and scientific expertise accompanies the occurrence of sinkholes in Florida, but compared to the study of geological hazards in other parts of the world, the study of Florida sinkholes is in its early stages. There is still a great deal we do not know. A tremendous amount of work was done on the subject in the 1980s and early 1990s by the Florida Sinkhole Research Institute at the University of Central Florida, but since then, unfortunately, there has not been a coordinated effort to study Florida sinkholes. Important sinkhole research still emerges each year, of course, but the state no longer has an institute that serves as a clearing house.

Recently, I visited the site of the Winter Park Sinkhole—the sinkhole event that triggered so much interest in sinkhole research in Florida. The site today is a tranquil park on the edge of a very busy street in Winter Park. There is little evidence of the disastrous event that caused so much property damage and so much public anxiety. The site contains a small lake—really a pond—surrounded by some urban development and a tree-lined public area. The site is a testament to human relience in the face of disaster and to the great skill of our engineers in stabilizing and reconstructing such areas. Yet looking back, sinkhole engineering and sinkhole study was in its early stages when that sinkhole formed. The Winter Park disaster reeled out so many threads of inquiry that today we have a well-developed body of literature and significant public policy approaches to sinkhole formation. While there is much left to do, we have come a long way in weaving a fabric of sinkhole knowledge.

Here is what we know. We know that a sinkhole forms from the solution of limestone in the subsurface. When the void grows to the point that the roof above the void can no longer be supported, some form of collapse occurs. The collapse can be sudden, forming a collapse sinkhole, or it can be slower, with sediment gradually raveling into subsurface voids to form a subsidence sinkhole. Sinkholes also form when bedrock at the surface dissolves to create a depression. These types of sinkholes are rare in Florida because bedrock is typically covered in most parts of the state by marine sediment.

We also know that there is a distinct geographic distribution of sinkholes and sinkhole types. Old sinkholes, called paleo-sinkholes, are most common on the ridges of Florida. These sinkholes formed tens of thousands of years ago when the sea level was higher and when water tables were higher on the peninsula. Sinkholes still occur on Florida's ridges, but many of the lakes and complex depressions that formed on the ridges are much older than sinkholes in the surrounding lowlands. Sinkholes that are found closer to sea level in the flatter portions of the peninsula are generally younger than the sinkholes found on the ridges. The lower areas were covered by seawater for longer periods of time. Seawater inhibits solution of limestone, so karst processes were not active in these lower areas during times of high sea level. On the other hand, karst processes were active in these lower areas between high sea level stands. There is evidence in these low areas of old sinkholes, but they are filled with marine sediment deposited during high water events.

We also know that sinkholes are rare where the sediment cover above bedrock is thick. For example, in the Tampa Bay area, sediment cover thickens as one travels from Tampa to Sarasota. In Sarasota County, sinkholes are rare events, whereas in Tampa, they are quite common. We also see this phenomenon in relatively small areas. In Hillsborough County, the thickness of sediment differs from place to place, and there is a general association between sinkhole formation and the sediment cover being thin. We also know that sinkholes are not particularly common where bedrock is young. In the Everglades, for example, where the bedrock at the surface is only a few thousands of years old, sinkholes are uncommon. However, where they are present, they are important for local hydrology and ecology.

The technology of sinkhole detection and mapping has increased greatly over the last twenty years. While it is neither easy nor inexpensive to detect void space underground, there are a variety of geophysical techniques that can be used to find subsurface variations in Florida. The most useful technique developed to date is ground-penetrating radar. Used widely in geotechnical investigations in Florida, GPR is an efficient, cost-effective approach that can find near-surface underground voids. The imaging and detection capabilities of GPR have increased greatly, and new advances in the field occur each year.

Likewise, improvements in sinkhole mapping are occurring. Prior to the use of Geographic Information Systems, sinkhole databases were not linked to any particular mapping software. Now, with such mapping programs, we can import geographic coordinates to pinpoint with great accuracy the location of individual sinkholes on an electronic map. The Florida Sinkhole Research Institute began mapping sinkhole occurrences in the 1980s, and the Florida Geological Survey continues that effort. Today, one can download sinkhole data from the FGS website easily.

Electronic mapping also makes the use of digital topographic maps particularly helpful for sinkhole studies. The USGS 1:24,000 quadrangles can be used to map depressions. In addition, aerial photos can be analyzed electronically to map depressions and circular karst features for analysis. The use of airborne laser swath mapping holds promise for detailed mapping in karst terrain, providing remarkable accuracy and eliminating many scale problems associated with both the use of standard USGS topographic maps and the interpretation of aerial photos.

Because of improved mapping capabilities and the maintenance of a statewide sinkhole database, we can broadly predict the areas where sinkholes are likely to occur, and because the geologic past can help us predict the future, we know that areas where sinkholes have commonly formed in the past are likely areas for sinkhole development today. These regional groupings help us understand why particular geologic conditions favor sinkhole development over other geologic conditions. We also know that many natural processes, such as the oxidation of organic matter and the expansion and contraction of shrink-swell clays, can mimic sinkhole development on the surface of the earth.

We have come a long way in developing sinkhole policy in the state. The most significant advance is the implementation of mandatory catastrophic collapse coverage for all homeowner insurance policies and optional sinkhole insurance. Florida is the only state that requires such coverage, and the impact of this requirement has been significant. For one thing, homeowners have saved millions of dollars since the implementation of the requirement. For another, new geotechnical expertise has developed to provide professional assessment of sinkholes, and new fields of law have emerged with the development of litigation around sinkhole insurance claims. In addition, communities have implemented land-use regulations to protect sinkhole watersheds and the overall sinkhole environment.

Florida's engineers and architects have also risen to the task of designing structures for our unstable landscape. They have developed innovative ways to protect projects such as homes and roads from collapse damage. They have found ways to stabilize the landscape prior to construction, and they have developed complex techniques for mitigating damage after a collapse occurs. They can fill subsurface voids to prevent further collapse, and they can reinforce homes and find ways to preserve structural integrity after a collapse occurs. Geotechnical advances are occurring all the time in response to the challenging problems of ground failure as a result of karst activity. A wise man once said that necessity is the mother of invention. As the instability of Florida's surface expands, the state's need for invention is great.

With all of the advances we have made, there are still many things we do not know about sinkholes. In fact, our sinkhole knowledge is really in its early stages of development. Students of the geology of many parts of the United States find a rich literature that dates back to the nineteenth century. This work reflects the growth of geologic thought. Some of this early literature is descriptive geology that defines the bedrock, structural geology, and geomorphology of a region. In the study of Florida sinkholes, we do not have such a rich literature. Indeed, the early literature on Florida geology as a whole is thin. While one could claim that the geology of Florida is not as dramatic or interesting as, say, the geology of New Mexico, the lack of fundamental research on the region is still unusual. It is not surprising that the literature on sinkholes balloons in the second half of the twentieth century as technology made it possible to glimpse into the subsurface to detect voids and to discern variations in the bedrock and structural geology. Along with these advances, we also had major advances in mapping. The growth of GIS and capabilities in computer analysis of remotely sensed images has enhanced our ability to conduct investigations. This technological growth has come about at a time when widespread growth and suburban sprawl drives the population of Florida to expand further into active karst areas. The major Winter Park sinkhole occurred in the midst of all this urban development and served as an impetus for future studies.

The Winter Park sinkhole, then, is a watershed event. Knowledge about sinkholes has grown more in the decades since its occurrence than in the decades prior to its formation. Even so, many research questions remain unanswered. Geologically, we are still trying to understand the age of sinkholes. Clearly there were periods when sinkholes actively form in particular places and periods when they do not form. What is responsible for this discrepancy? Is it related solely to sea-level variations, or is there a climatic factor, separate from the sea level issue, that is responsible? How closely tied are sinkhole formation and speleogenesis? Can we find a way to map or classify sinkholes by age of formation? Also, can we identify and map sinkholes that filled as a result of marine transgressions over the state? Can they be separated from those that filled as a result of colluvial and slope processes? The dating of sinkhole formation and the mapping of filled sinkholes will certainly give us a lot of information about the geological and environmental history of the state. Although there are efforts under way in some areas to map sinkholes that were destroyed as a result of urbanization, can we conduct a statewide survey of these types of sinkholes in order to better understand the pre-development distribution of them on the landscape?

Along with the unknowns surrounding sinkhole age, there is also much we need to learn about sinkhole formation. Few sinkholes are ever studied in detail. In fact, only those that cause some type of damage are regularly evaluated as part of a formal insurance investigation. These evaluations typically go into the gray literature and, because they are proprietary, seldom enter the public domain. Making such reports more available to researchers would enable them to better understand the characteristics of sinkhole formation as well as their spatial variation. Efforts have been made to map known subsurface voids in the United Kingdom. Unfortunately, we do not have any similar comprehensive effort under way in Florida. It would be fascinating to collate all of the information that exists on the distribution of known subsurface voids in Florida. The database would have to be three-dimensional, as voids can be located above and beneath one another because of the Swiss cheese– or sponge-like nature of the underlying limestone. The development of such a database would assist land-use managers and would further our understanding of the nature of karst in Florida.

Point of collapse is another dimension of sinkhole knowledge we need to expand. While we know that human activities can trigger sinkhole formation, we do not regularly measure the point at which the surface is apt to collapse. For example, we know that excessive withdrawal of water from the aquifer can cause surface collapse, but we do not know exactly when that collapse will occur. Can we develop models to predict sinkhole collapse in various parts of Florida where induced sinkholes have occurred? While such prediction is not yet cost effective, primary research could be conducted to assess how to develop such models. Several types of induced sinkholes could be studied: those in agricultural fields caused by pumping and watering; those that form with the addition of storm water weight; and those caused by housing construction.

We also need to better understand the mechanism of sinkhole formation. We have a variety of models of how sinkholes form: collapse, solution, and subsidence. While all three provide clear frameworks for the formation of sinkholes, much more needs to be understood about their nuances. For example, how do void spaces change with time? At what rate does raveling occur? How does water chemistry influence the formation of solution sinkholes? There are many more questions that can be asked, but the point is the same: we need to build upon the excellent models of sinkhole formation that exist. Along with this, we need to better understand how the aquifer systems in Florida work. A great deal of research has been conducted in this field, but we need to continue to develop better hydrogeologic models as new knowledge emerges.

We also need to develop better risk models of sinkhole formation. As noted above, we have made strong advances in the development of sinkhole databases in Florida, but, to date, we have not developed a very sophisticated risk model. There are certainly broad regions that have been mapped that associate sinkhole formation with things like depth to bedrock. However, the associations are on a regional scale and do not account for the finer variations that exist within the regions. The reluctance to develop more granular risk models and maps may be due to a fear of insurance redlining and of a loss of property value. However, I believe that the development of a detailed sinkhole risk map could further the science of sinkholes in Florida. Such a map could be refined as we collect more information on, say, the distribution of underground voids, and they could be useful in assessing appropriate site preparation prior to development.

We also need to have a statewide discussion that includes various stakeholders' input on sinkhole policy in the state. The state has been progressive both in requiring catastrophic collapse insurance for all homeowner policies and in requiring insurers to offer sinkhole insurance. But the state definition of the term "sinkhole" in the law requiring insurance, as it is currently written, needs clarification. The current law leaves a great deal of ambiguity in the definition and leads to quite a variation of opinion in the professional community as to the interpretation of the term. As a result, there are often challenges to geotechnical assessments conducted on behalf of homeowners or insurance companies when a sinkhole damage claim is made. The problem is compounded by the fact that other natural processes can mimic sinkhole damage. Can the law be rewritten to clarify the term "sinkhole"? Or will rewriting the law lead to further complication? Also, what about the other natural processes? Is it possible to develop a plan to cover things such as oxidizing organic matter or surface expansion and contraction as a result of the presence of shrink-swell clays? There also needs to be more clarity about why homes are denied sinkhole coverage.

Along with these insurance issues, it might be useful to have a statewide discussion on the current policy regarding sinkhole-damaged homes. Would it be useful to develop a statewide, public database of homes damaged by sinkholes? In some instances, insufficient repairs are made after a sinkhole damage claim is paid. Is it possible to better evaluate sinkhole repairs after a claim is paid? We need to better understand the impact of the various sinkhole insurance rules on the public, on the state, and on the insurance industry. Although no one wants to create a situation in which insurance companies pull out of the region, as happened after Hurricane Andrew, we need to continue to evaluate how to protect the public investment in property and how to protect neighborhoods from declining property values as a result of sinkhole formation.

We need to evaluate other public policies related to sinkholes as well, especially those involving land-use rules and environmental regulations. Many parts of the country have adopted land-use measures that specifically take sinkhole risk into consideration. Improved site development is required in high-risk areas, and homeowners should be made aware of the types of risks they are taking when moving into such areas. Also, many parts of the country have vigorous setback rules that require that development be placed some distance from karst features. While some areas of Florida have such rules, there is little uniformity in the state. We also need to evaluate how stormwater drainage into sinkholes impacts the aquifer. We know that sinkholes are typically no longer used for storm-water drainage, but many drainage systems do directly feed into sinkholes, and we do not fully understand how the wastewater impacts our aquifer.

In recent years, a great deal of attention has been focused on the impact of karst on large projects such as the Crosstown Expressway in Tampa and large high-rises in Orlando. Many potential problems can be managed by thorough analysis of sites prior to development. But in some instances, it seems that current approaches to site assessment are not sufficient. Also, many structures, such as the previously discussed large gypsum stack, were built prior to the ability to assess the potential for karst formation. These types of projects, which were developed prior to advances in, say, GPR technology, need to be reevaluated in order to protect the public. While this may seem a costly, and in some individuals' minds, unnecessary, step, to leave these large structures unevaluated is like whistling in a graveyard. Given the environmental disaster that occurred when the sinkhole formed in the gypsum stack, it seems prudent to avoid other disasters that could occur on other important large structures.

We also need to continue to improve our ability to detect sinkholes. While GPR is a wonderful tool, we should continue to strive for improved ways to assess individual properties for the presence of sinkholes. Perhaps one day, we can develop a relatively inexpensive tool and make it available to the public so that people can assess their own property or property they would like to purchase. While this may not seem possible in the short term, our technology has become so much more democratic in recent years that it would not be surprising if real estate agents or homebuyers conducted some type of sinkhole evaluation on their own in the near future—maybe using an app on a computer or other device. Only a few years ago, GPS was available only to very few subscribers. Can we develop GPR in a way that members of the general public can interpret sinkhole risks on their own?

We have a long way to go in sinkhole research in Florida. I am certain that a concentrated effort in enhancing sinkhole research will benefit the state's people. There are many research groups in Florida that focus on sinkhole research as part of their mission, though currently, there is little coordination of these efforts because there is so much research to be done! However, I hope that in the future we can return to the days of the 1980s and early 1990s, when there were better-coordinated efforts in the state. There are a number of different groups that could coordinate such an effort. Certainly, the Florida Geological Survey conducts a tremendous amount of valuable research and could manage this coordination as part of its mission. Also, one of the various university research groups could take on this effort. I hope this book provides a springboard for these discussions. I also humbly believe that this book can begin a new phase of sinkhole research in Florida that coordinates the efforts of our very talented experts. While there is much to do, we must also recognize that much has been done. Many researchers have built a solid body of research on sinkholes in Florida, and I dedicate this book to them and the people of Florida.

References

Amaleh, L. J., J. D. Grob, R. H. Gorny. 1993. Ground stabilization for foundation and excavation construction in Florida karst topography. *Environmental Geology* 22:308–13.

Anderson, L. 2004. *Paynes Prairie: The great savanna. A history and guide.* Sarasota, Fla.: Pineapple.

Armentano, T. V., D. T. Jones, M. S. Ross, and B. W. Gamble. 2002. Vegetation pattern and process in tree islands of the southern Everglades and adjacent areas. In *Tree islands of the Everglades*, ed. F. Sklar and A. Valk, 225–81. Heidelberg: Springer.

Arrington, D. V., and R. C. Lindquist. 1987. Thickly mantled karst of the Interlachen, Florida area. In *Karst hydrogeology: Engineering and environmental applications*, ed. B. F. Beck and W. L. Wilson, 31–39. Rotterdam: A. A. Balkema.

Arthur, J. D. 1991. *The geomorphology, geology, and hydrogeology of Lafayette County.* Open File Report no. 46. Tallahassee: Florida Geological Survey.

Bahtijarevic, A. 1989. Sinkhole density of the Forest City Quadrangle. In *Engineering and environmental impacts of sinkholes and karst: Proceedings of the Third Multidisciplinary Conference on Sinkholes and the Engineering and Environmental Impacts of Karst*, ed. B. F. Beck, 75–82. Rotterdam: A. A. Balkema.

Barco, J. W. 2001. A soil toposequence of the southern portion of the Brooksville Ridge. Master's thesis, University of South Florida, Tampa.

Bartram, W. 1792. *Travels through North and South Carolina, Georgia, east and west Florida, the Cherokee country, the extensive territories of the Muscogulges or Creek Confederacy, and the country of the Chactaws.* Philadelphia: James and Johnson.

Beck, B. F. 1986. A generalized genetic framework for the development of sinkhole and karst in Florida, USA. *Environmental Geology and Water Sciences* 8 (1/2): 5–18.

Beck, B. F., D. Bloomberg, J. T. Trommer, and K. McDonald. n.d. *A field guide to some illustrative karst features in the Tampa area, Hillsborough County, Florida.* Report no. 89-90-1. Orlando: Florida Sinkhole Research Institute.

Beck, B. F., and S. Sayed. 1991. *The sinkhole hazard in Pinellas County: A geologic summary for planning purposes.* Orlando: Florida Sinkhole Research Institute.

Belesky, R. M., R. H. Hardy, and F. F Strouse. 1987. Sinkholes in airport pavements: Engineering implications. In *Karst hydrology: Engineering and environmental applications: Proceedings of the second multidisciplinary conference on sinkholes and the engineering and environmental impacts of karst*, ed. B. F. Beck and W. L. Wilson, 411–17. Rotterdam: A. A. Balkema.

Bengtsson, T. O. 1987. The hydrologic effects from intense ground-water pumpage in east-central Hillsborough County, Florida. In *Karst hydrology: Engineering and environmental applications: Proceedings of the second multidisciplinary conference on sinkholes and the engineering and environmental impacts of karst*, ed. B. F. Beck and W. L. Wilson, 109–14. Rotterdam: A. A. Balkema.

Benson, R. C., L. Yuhr, and P. Passe. 1995. Assessment of potential karst conditions for a new bridge in the Florida Keys. In *Proceedings of the Symposium on the Application of Geophysics to Engineering and Environmental Problems*, ed. R. S. Bell, 529–39. Orlando, Fla.: Environmental and Engineering Geophysical Society.

Bergado, T. D., C. Areepitak, and F. Prinzl. 1984. Foundation problems on karstic limestone formations in western Thailand—A case of Khao Laem Dam. In *Sinkholes: Their geology, engineering and environmental impact: Proceedings of the First Multidisciplinary Conference on Sinkholes and the Engineering and Environmental Impacts of Karst, Orlando, Florida, 15-17 October 1984*, ed. B. F. Beck, 387-401. Rotterdam: A. A. Balkema.

Bloomberg, D., S. B. Upchurch, M. L. Hayden, and R. C. Williams. 1987. Cone-penetrometer exploration of sinkholes: Stratigraphy and soil properties. In *Karst hydrogeology: Engineering and environmental applications*, ed. B. F. Beck and W. L. Wilson, 145–51. Rotterdam: A. A. Balkema.

Bonaparte, R., and R. R. Berg. 1987. The use of geosynthetics to support roadways over sinkhole prone areas. In *Karst Hyrdrology: Engineering and Environmental Applications: Proceedings of the Second Multidisciplinary Conference on Sinkholes and the Engineering and Environmental Impacts of Karst*, ed. B. F. and W. L. Wilson, 437-45. Rotterdam: A. A. Balkema.

Borel, J. S. 1997. Coevolution of landscape and culture: The vegetation of Indian shell mounds in Florida's Ten Thousand Islands. Master's thesis, University of Florida, Gainesville.

Brandt, L. A., J. E. Silveira, and W. M. Kitchens. 2002. Tree islands of the Arthur R. Marshall Loxahatchee National Wildlife Refuge. In *Tree islands of the Everglades*, ed. F. H. Sklar and A. Van Der Valk, 311–35. Heidelberg: Springer.

Brinkmann, R., and P. Reeder. 1994. The influence of sea-level change and geologic structure on cave development in west-central Florida. *Physical Geography* 15 (1):52–61.

———. 1995. The relationship between surface soils and cave sediments: An example from west-central Florida, USA. *Cave and Karst Science* 22:95–102.

Brinkmann, R., K. Wilson, N. Elko, L. D. Seale, L. J. Florea, and H. L. Vacher. 2007. Sinkhole distribution based on pre-development mapping in urbanized Pinellas County, Florida, USA. In *Natural and anthropogenic hazards in karst areas: Recognition, analysis and mitigation* ed. M. Parise and J. Gunn, 5–11. Special publication 279. Geological Society of London.

Carr, R. S. 2002. The archaeology of Everglades tree islands. In In *Tree islands of the Everglades*, ed. F. H. Sklar and A. Van Der Valk, 187–206. Heidelberg: Springer.

Carr, W. J., and D. C. Alverson. 1959. *Stratigraphy of Middle Tertiary rocks in part of west-central Florida*. U.S. Geological Survey Bulletin no. 1092. Washington, DC: U.S. Government Printing Office.

Carson, R. 1962. *Silent spring*. Boston: Houghton Mifflin.

Connor, W. H., T. W. Doyle, and D. Mason. 2002. Water depth tolerances of dominant tree island species: What do we know? In *Tree islands of the Everglades*, ed. F. H. Sklar and A. Van Der Valk, 207–23. Heidelberg: Springer.

Conyers, L. B., and D. Goodman. 1997. *Ground penetrating radar: An introduction for archaeologists*. Walnut Creek: AltaMira Press.

Cooke, C. W. 1939. Scenery of Florida interpreted by a geologist. Bulletin no. 17. Florida Geological Survey.

Currin, J. L., and B. L. Barfus. 1989. Sinkhole distribution and characteristics in Pasco County, Florida. In *Engineering and environmental impacts of sinkholes and karst: Proceedings of the*

Third Multidisciplinary Conference on Sinkholes and the Engineering and Environmental Impacts of Karst, ed. B. F. Beck, 97–106. Rotterdam: A. A. Balkema.

Daniels, D. J. 1996. *Surface penetrating radar*. Exeter: Short Run Press.

Davis, W. M. 1899. The geographic cycle. *Geographic Journal* 14(5):481–504.

Day, M. J. 2004. Karstic problems in the construction of Milwaukee's Deep Tunnels. *Environmental Geology* 45(6): 859-63.

Dinger, J. S., and J. R. Rebmann. 1986. Ordinance for the control of urban development in sinkhole areas in the Blue Grass karst region, Lexington, Kentucky. In *Proceedings of the Conference on Environmental Problems in Karst Terrains and Their Solutions*, 163–80. Dublin, Ohio: National Water Well Association.

Dinger, J. S., D. P. Zourarakis, and J. C. Currens. 2007. Spectral enhancement and automated extraction of potential sinkholes from NAIP Imagery—Initial investigations. *Journal of Environmental Informatics* 10(1):22–29.

Douglas, M. S. 1947. *The Everglades: River of grass*. New York: Rinehart.

DuBar, J. R. 1991. Florida peninsula. In Quaternary geology of the Gulf of Mexico Coastal Plain, 595–604, chpt. 19 of *Quaternary Nonglacial Geology: Conterminous U.S.*, ed. R. B. Morrison. Vol. K2, Geology of North American series. Geological Society of North America.

Eastman, K. L., A. M. Butler, and C. C. Lilly. 1995. The effect of mandating sinkhole coverage in Florida homeowners insurance policies. *CPCU Journal* 48(3):165–76.

Edmonds, C. N., C. P. Green, and I. E. Higginbottom. 1987. *Subsidence hazard prediction for limestone terrains as applied to the English Cretaceous Chalk*. Special publication. Geological Society of London.

Engineering News Record. 1996. NYC Plugs Sewer Sinkhole. Vol. 236, 4.

Florea L. J., and H. L. Vacher. 2005. Springflow hydrographs: Eogenetic vs. telogenetic karst. *Groundwater* 44(3):352–61.

Florea, L. J., H. L. Vacher, B. Donahue, B. and D. Naar. 2007. Quaternary cave levels in peninsular Florida. *Quaternary Science Reviews* 26(9–10): 1344–1361.

Florida Geological Survey (FGS). 1997. *Geologic and geotechnical assessment for the evaluation of sinkhole claims*. Open File Report no. 72. Tallahassee: FGS.

———. 2011. Subsidence incidence reports. Florida Department of Environmental Protection website, http://www.dep.state.fl.us/geology/gisdatamaps/SIRs_database.htm/.

Florida Lake Management Society, and Florida Lakewatch. 2011. *Florida Atlas of Lakes*. http://www.wateratlas.usf.edu/AtlasOfLakes/Florida/.

Ford, D. C., and P. W. Williams. 1989. *Karst hydrology and geomorphology*. Chichester, UK: John Wiley and Sons.

Foshee, J., and B. Bixler. 1994. Cover-subsidence sinkhole evaluation of State Road 434, Longwood, Florida. *Journal of Geotechnical Engineering* 120(11): 2026-2040.

Frank, E. F., and B. F. Beck. 1991. *An analysis of the cause of subsidence damage in the Dunedin, Florida area 1990/1991*. Orlando: Florida Sinkhole Research Institute.

Fretwell, J. D. 1988. *Water resources and effects of ground-water development in Pasco County, Florida*. Water Resources Investigations Report no. 87-4188. Tallahassee: U.S. Geological Survey.

Fuleihan, N. F., J. E. Cameron, and J. F. Henry. 1997. The hole story: How a sinkhole in a phosphogypsum pile was explored and remediated. In *The engineering geology and hydrogeology of karst terranes: Proceedings of the Sixth Multidisciplinary Conference on Sinkholes and the Engineering and Environmental Impacts of Karst, Springfield Missouri, 6-9 April, 1997*, ed. B. F. Beck and J. B. Stephenson, 363–70. Rotterdam: A.A. Balkema.

Gaines, M. S., C. R. Sasso, J. E. Diffendorfer, and H. Beck. 2002. Effects of tree island size and water on the population dynamics of small mammals in the Everglades. In *Tree islands of the Everglades*, ed. F. H. Sklar and A. Van Der Valk, 429–44. Heidelberg: Springer.

Gao, Y., R. G. Tipping, and E. C. Alexander. 2006. Applications of GIS and database technologies to manage a karst feature database. *Journal of Cave and Karst Studies* 68(3):144–52.

Garlanger, J. E. 1984. Remedial measures associated with sinkhole-related foundation distress. In *Sinkholes: Their geology, engineering and environmental impact: Proceedings of the First Multidisciplinary Conference on Sinkholes and the Engineering and Environmental Impacts of Karst, Orlando, Florida, 15–17 October 1984*, ed. B. F. Beck, 335–41. Rotterdam: A. A. Balkema.

Gawlik, D. E., P. Gronemeyer, and R. A. Powell. 2002. Habitat use patterns of avian seed dispersers in the central Everglades. In *Tree islands of the Everglades*, ed. F. H. Sklar and A. Van Der Valk, 445–68. Heidelberg: Springer.

Gilboy, A. E. 1987. Ground penetrating radar: Its application in the identification of subsurface solution features—A case study in west-central Florida. In *Karst hydrogeology: Engineering and environmental applications*, ed. B. F. Beck and W. L. Wilson, 197–203. Rotterdam: A. A. Balkema.

Goehring, R. L., and S. M. Sayed. 1989. Mann road sinkhole stabilization: A case history. In *Engineering and environmental impacts of sinkholes and karst: Proceedings of the Third Multidisciplinary Conference on Sinkholes and the Engineering and Environmental Impacts of Karst*, ed. B. F. Beck, 319–25. Rotterdam: A. A. Balkema.

Gordon, M. M. 1987. Sinkhole repair: The Bottom Line. In *Karst hydrology: Engineering and environmental applications: Proceedings of the second multidisciplinary conference on sinkholes and the engineering and environmental impacts of karst*, ed. B. F. Beck and W. L. Wilson, 419–24. Rotterdam: A. A. Balkema.

Gunderson, L. H., and L. L. Loope. 1982. *A survey and inventory of the plant communities in the Pinecrest area, Big Cypress National Preserve*. South Florida Research Center Report no. T-655. Homestead, Fla.: Everglades National Park.

Gutiérrez, F., J. Guerrero, and P. Lucha. 2008. A genetic classification of sinkholes illustrated from evaporite paleokarst exposures in Spain. *Environmental Geology* 53(5):993–1006.

Hansen, B. C. S., E. C. Grimm, and W. A. Watts. 2001. Palynology of the Peace Creek site, Polk County, Florida. *Geological Society of America* 113:682–92.

Harvey, J. W., S. L. Krupa, C. Gefvert, R. H. Mooney, J. Choi, S. A. King, and J. B. Giddings. 2002. *Interactions between surface water and ground water and effects on mercury transport in the north-central Everglades*. Water Resources Investigations Report no. 02-4050. Tallahassee: U.S. Geological Survey.

Heisler, L., D. T. Towles, L. A. Brandt, and R. T. Pace. 2002. Tree island vegetation and water management in the central Everglades. In *Tree islands of the Everglades*, ed. F. H. Sklar and A. Van Der Valk, 283–309. Heidelberg: Springer.

Henry, J. F. 1989. Ground modification techniques applied to sinkhole remediation. In *Karst Hyrdrology: Engineering and Environmental Impacts of Sinkholes and Karst: Proceedings of the Third Multidisciplinary Conference on Sinkholes and the Engineering and Environmental Impacts of Karst*, ed. B. F. Beck, 327–32. Rotterdam: A. A. Balkema.

Hildebrand, P. B., and R. Oros. 1987. Wastewater reuse in karst terrain—A case study. In *Karst hydrogeology: Engineering and environmental applications*, ed. B. F. Beck and W. L. Wilson, 259. Rotterdam: A. A. Balkema.

Hoenstine, R. W., E. Lane, S. M. Spencer, and T. O'Carroll. 1987. A landfill site in a karst environment, Madison County, Florida—A case study. In *Karst hydrogeology: Engineering and environmental applications*, ed. B. F. Beck and W. L. Wilson, 253–58. Rotterdam: A. A. Balkema.

Hollingshead, J. J. 1984. A contour map, volume estimate, and description of Teague's Sinkhole. In *Sinkholes: Their geology, engineering and environmental impact: Proceedings of the First Multidisciplinary Conference on Sinkholes and the Engineering and Environmental Impacts of Karst, Orlando, Florida, 15–17 October 1984*, ed. B. F. Beck, 105–9. Rotterdam: A. A. Balkema.

Holszchuh, J. C. 1972. State of water resources within the Peace River basin with emphasis on ground water. Technical report to the Southwest Water Management District.

Horvitz, C. C., S. McMann, and A. Freedman. 1995. Exotics and hurricane damage in three hardwood hammocks in Dade County parks. *Florida Journal of Coastal Research* S1(21):145–58.

Hurston, Z. N. 1937. *Their eyes were watching God*. Philadelphia: J. P. Lippincott.

Hyatt, J. A., and P. M. Jacobs. 1996. Distribution and morphology of sinkholes triggered by flooding following Tropical Storm Albertoat Albany, Georgia, USA. *Geomorphology* 17(4):305–16.

International Union of Speleology (UIS) and International Association of Hydrogeologists (IAH). 2011. *Speleogenesis Scientific Network*. Website. Glossary of cave and karst terms. http://network.speleogenesis.info/directory/glossary/index.php/.

Jammal, S. E. 1984. Maturation of the Winter Park sinkhole. In *Sinkholes: Their geology, engineering and environmental impact: Proceedings of the First Multidisciplinary Conference on Sinkholes, Orlando, Florida, 15–17 October 1984*, ed. B. F. Beck, 363–69. Rotterdam: A. A. Balkema.

Jammal and Associates, Inc. 1982. *The Winter Park sinkhole: A report of the investigation, findings, and recommendations*. Winter Park, Fla.: Winter Park City Commission.

Jensen, J. H. 1987. Valley poljes in Florida karst. In *Karst hydrogeology: Engineering and environmental applications*, ed. B. F. Beck and W. L. Wilson, 31–39. Rotterdam: A. A. Balkema.

Keddy, P. A. 2000. *Wetland ecology: Principles and conservation*. Cambridge, UK: Cambridge University Press.

Kemmerly, P. R. 1993. Sinkhole hazards and risk assessment in a planning context. *Journal of the American Planning Association* 592: 221–29.

Kimrey, J. O. 1978. *Preliminary appraisal of the geohydrological aspects of drainage wells, Orlando area, central Florida*. Water Resources Investigations Report no. 78-37. Tallahassee: U.S. Geological Survey.

Kindinger, J. L., J. B. Davis, and J. G. Flocks. 1999. Geology and evolution of lakes in north-central Florida. *Environmental Geology* 38(4):301–21.

Kirby, R. K., C. H. Hobbs, and A. J. Mehta. 1994. Shallow stratigraphy of Lake Okeechobee, Florida: A preliminary reconnaissance. *Journal of Coastal Research* 10(2):339–50.

Klimchouk, A. B. 2005. Subsidence hazards in different types of karst: Evolutionary and speleogenetic approach. *Environmental Geology* 48:287–96.

———. 2007. *Hypogene speleogenesis: Hydrological and morphogenetic perspective*. Carlsbad, N.M.: National Cave and Karst Research Institute.

Klimchouk, A. B., D. C. Ford, A. N. Palmer, and W. Dreybrodt. 2000. *Speleogenesis: Evolution of karst aquifers*. Huntsville, Ala.: National Speleological Society.

Klimchouk, A. B., and D. Lowe, eds. 2002. Implication of speleological studies for karst subsidence hazard assessment. *Journal of Speleology* 31(theme issue).

Kuhns, G. L., L. M. Phelps, P. Bryant, M. Cox, and E. A. Cox. 1987. Subsurface indicators of potential sinkhole activity at the Maitland Colonnades project in Maitland, Florida. In *Karst hydrogeology: Engineering and environmental application Proceedings of the Second Multidisciplinary Conference on Sinkholes and the Engineering and Environmental Impacts of Karst*, ed. B. F. Beck and W. L. Wilson, 365–81. Rotterdam: A. A. Balkema.

Kuhns, G. L., L. M. Phelps, B. P. Marshall, and E. A. Cox. 1987. Subsurface indicators of potential sinkhole activity at the Maitland Colonnades project in Maitland, Florida. In *Karst hydrology: Engineering and environmental applications: Proceedings of the second multidisciplinary conference on sinkholes and the engineering and environmental impacts of karst*, ed. B. F. Beck and W. L. Wilson, 365–81. Rotterdam: A.A. Balkema.

Land, L. A., and C. K. Paull. 2000. Submarine karst belt rimming the continental slope in the Straights of Florida. *Geo-Marine Letters* 20(2):123–32.

Land, L. A., C. K. Paull, and B. Hobson. 1995. Genesis of a submarine shallow sinkhole without subaerial exposure: Straights of Florida. *Geology* 23(10):949–51.

Lane, E. 1993. Subsurface karst features in Florida. In *Applied karst geology*, ed. B. F. Beck, 199–204. New York: Taylor and Francis.

Light, S. S., and J. W. Dineen. 1994. Water control in the Everglades: A historical perspective. In *Everglades: The ecosystem and its restoration*, ed. S. Davis and J. Ogden, 47–84. Delray Beach, Fla.: St. Lucie Press.

Limestone Resource Committee of the North Jersey Resource Conservation and Development Council. 1993. A karst model ordinance. In *Applied karst geology*, ed. B. F. Beck, 273–76. New York: Taylor and Francis.

Littlefield, J. R., M. A. Culbreth, S. B. Upchurch, and M. T. Stewart. 1984. Relationship of modern sinkhole development to large scale photolinear features. In *Sinkholes: Their geology, engineering and environmental impact: Proceedings of the First Multidisciplinary Conference on Sinkholes and the Engineering and Environmental Impacts of Karst, Orlando, Florida, 15–17 October 1984*, ed. B. F. Beck, 189–96. Rotterdam: A. A. Balkema.

Lodge, T. E. 1994. *The Everglades handbook: Understanding the ecosystem*. Delray Beach, Fla.: St. Lucie Press.

Loope, L. L., and N. H. Urban. 1980. *A survey of fire history and impact in tropical hardwood hammocks in the east Everglades and adjacent portions of Everglades National Park*. South Florida Research Center Report no. T-592. Homestead, Fla.: National Park Service.

Louw, J. M., P. H. Goedhart, and F. J. van Zyl. 1984. A model study of a proposed concrete road pavement over a potential sinkhole area. In *Sinkholes: Their geology, engineering and environmental impact: Proceedings of the First Multidisciplinary Conference on Sinkholes and the Engineering and Environmental Impacts of Karst, Orlando, Florida, 15–17 October 1984*, ed. B. F. Beck, 391–96. Rotterdam: A. A. Balkema.

Manda, A. K., and M. R. Gross. 2006. Identifying and characterizing solution conduits in karst aquifers through geospatial GIS analysis of porosity from borehole imagery: An example from the Biscayne Aquifer, south Florida USA. *Advances in Water Resources* 29(3):383–96.

Marshall, A. R. 1982. *For the future of Florida: Repair the Everglades*. Coconut Grove, Fla.: Friends of the Everglades.

Mason, D. H., and A. Van Der Valk. 2002. Vegetation, peat elevation and peat depth on two tree islands in Water Conservation Area 3-A. In *Tree islands of the Everglades*, ed. F. H. Sklar and A. Van Der Valk, 337–56. Heidelberg: Springer.

McCarty, K. 2004. *Long range program plan, 2005–2010*. Tallahassee:Office of Insurance Regulation.

Menke, C. G., E. W. Meredith, and W. S. Wetterhall. 1961. *Water resources of Hillsborough County, Florida*. Report of Investigations no. 50. Tallahassee: Florida Geological Survey.

Meshaka, W. E., Jr., R. Snow, O. L. Bass Jr., and W. B. Robertson Jr. 2002. Occurrence of wildlife on tree islands in the southern Everglades. In *Tree islands of the Everglades*, ed. F. H. Sklar and A. Van Der Valk, 391–427. Heidelberg: Springer.

Metcalfe, S. J., and L. E. Hall. 1984. Sinkhole collapse induced by groundwater pumpage for freeze protection irrigation near Dover, Florida, January 1977. In *Sinkholes: Their geology, engineering and environmental impact: Proceedings of the First Multidisciplinary Conference on Sinkholes and the Engineering and Environmental Impacts of Karst, Orlando, Florida, 15–17 October 1984*, ed. B. F. Beck, 29–33. Rotterdam: A. A. Balkema.

Milanich, J. T. 1994. *Archaeology of Precolumbian Florida*. Gainesville: University Press of Florida.

Miller, J. A. 1997. Hydrogeology of Florida. In *The geology of Florida*, ed. A. F. Randazzo and D. S. Jones, 69–88. Gainesville: University Press of Florida.

Mitsch, W. J., and J. G. Gosselink. 1986. *Wetlands*. New York: Van Nostrand Reinhold.

Moore, H. L. 1984. Geotechnical considerations in the location, design, and construction of highways in karst terrain—"The Pellissippi Parkway extention," Knox-Blount Counties, Tennessee. In *Sinkholes: Their geology, engineering and environmental impact: Proceedings of the First Multidisciplinary Conference on Sinkholes and the Engineering and Environmental Impacts of Karst, Orlando, Florida, 15–17 October 1984*, ed. B. F. Beck, 385–89. Rotterdam: A. A. Balkema.

National Research Council. 2003. *Science and the Greater Everglades ecosystem restoration: An assessment of the Critical Ecosystem Studies Initiative*. Washington, DC: National Academies Press.

North American Commission on Stratigraphic Nomenclature. 2005. North American Stratigraphic Code. U.S. Geological Survey National Geologic Map Database. Website. http://ngmdb.usgs.gov/Info/NACSN/Code2/code2.html/.

Oberbauer, S. F., and S. Koptur. 1995. *Short and long term responses of non-tidal forest communities in Everglades National Park to Hurricane Andrew*. Report to Everglades National Park. Homestead, Fla.: National Park Service.

Olmsted, I. C., L. L. Loope, and C. E. Hilsenbeck. 1980. *Tropical hardwood hammocks of the interior of Everglades National Park*. South Florida Research Center Report no. T-604. Homestead, Fla.: National Park Service.

Olmsted, I. C., L. L. Loope, and R. P. Russell. 1981. *Vegetation of the southern coastal region of Everglades National Park between Flamingo and Joe Bay*. South Florida Research Center Report no. T-620. Homestead, Fla.: National Park Service.

Orem, W. H., D. A. Willard, H. E. Lerch, A. L. Bates, A. Boylan, and M. Comm. 2002. Nutrient geochemistry of sediments from two tree islands in Water Conservation Area 3B, the Everglades, Florida. In *Tree islands of the Everglades*, ed. F. H. Sklar and A. Van Der Valk, 153–86. Heidelberg: Springer.

Osmond, J. K., J. P. May, and W. F. Tanner. 1970. Age of the Cape Kennedy barrier and lagoon complex. *Journal of Geophysical Research* 75(2):469–79.

Padgett, D. A. 1993. Remote sensing application for identifying areas of vulnerable hydrogeology and potential sinkhole collapse within highway transportation corridors. In *Applied karst geology*, ed. B. F. Beck, 285–90. New York: Taylor and Francis.

Palmer, A. N. 2007. *Cave geology*. Trenton, N.J.: Cave Books.

Patton, T. H., and J. Klein. 1989. Sinkhole formation and its effect on Peace River hydrology.

In *Engineering and environmental impacts of sinkholes and karst: Proceedings of the Third Multidisciplinary Conference on Sinkholes and the Engineering and Environmental Impacts of Karst*, ed. B. F. Beck, 25–31. Rotterdam: A. A. Balkema.

Paukstys, B., A. H. Cooper, and J. Arustiene. 1999. Planning for gypsum geohazards in Lithuania and England. *Engineering Geology* 52(1–2):93–103.

Perkins, R. D. 1977. Depositional framework of Pleistocene rocks in south Florida. In *Quaternary sedimentation in south Florida: Geological Society of America Memoir No. 147*, ed. P. Enos and R. D. Perkins, 131–98. Geological Society of America.

Petuch, E. J. 1992. The Pliocene pseudoatoll of southern Florida and its associated gastropod fauna. In *Plio-Pleistocene stratigraphy and paleontology of southern Florida*, Special Publication no. 3, ed. T. M. Scott and W. D. Allmon, 101–16. Tallahassee: Florida Geological Survey.

Price, D. J. 1984. Karst progression. In *Sinkholes: Their geology, engineering and environmental impact: Proceedings of the First Multidisciplinary Conference on Sinkholes and the Engineering and Environmental Impacts of Karst, Orlando, Florida, 15–17 October 1984*, ed. B. F. Beck, 17–22. Rotterdam: A. A. Balkema.

———. 1989. Anatomy of a hazardous waste site in a paleosink basin. In *Engineering and environmental impacts of sinkholes and karst: Proceedings of the Third Multidisciplinary Conference on Sinkholes and the Engineering and Environmental Impacts of Karst*, ed. B. F. Beck, 189–96. Rotterdam: A. A. Balkema.

Puri, H. S., and R. O. Vernon. 1964. *Summary of the geology of Florida and a guidebook to the classic exposures*. Special Publication no. 5. Tallahassee: Florida Geological Survey.

Quinlan, J. F. 1986. Legal aspects of sinkhole development and flooding in karst terrains: 1. Review and synthesis. *Environmental Geology* 8(1):41–61.

Raghu, D. 1987. Determination of pile lengths and proofing of the bearing stratum of piles in cavernous carbonate formations. In *Karst hydrology: Engineering and environmental applications: Proceedings of the second multidisciplinary conference on sinkholes and the engineering and environmental impacts of karst*, ed. B. F. Beck and W. L. Wilson, 397–402. Rotterdam: A. A. Balkema.

Randazzo, A. F., and D. S. Jones. . 1997. *The geology of Florida*. Gainesville: University Press of Florida.

Reeder, P., and R. Brinkmann. 1998. Paleoenvironmental reconstruction of an Oligocene-aged island remnant in Florida, USA. *Cave and Karst Science* 25(1):7–13.

Reimus, R. 1997. *A list of species observed at Johnson Mound—An Indian midden near the Gulf Coast*. Report to Everglades National Park. Homestead, Fla.: National Park Service.

Rosenau, J. C., G. L. Faulkner, C. W. Hendry, R. W. Hull. 1977. *Springs of Florida*. Bulletin 31 (revised). Florida Bureau of Geology and Florida Department of Environmental Regulation.

Ross, M. S., J. F. Meeder, J. P. Sah, P. L. Ruiz, and G. Telesnicki. 1996. *The southeast saline Everglades revisited: A half-century of coastal vegetation change*. Final Report for Contract C-4244. Miami: Florida International University.

Ross, M. S., J. J. O'Brien, and L. J. Flynn. 1992. Ecological site characterization of Florida Keys terrestrial habitats. *Biotropica* 24(4):488–502.

Roux, P. 1987. The engineering-geological evaluation of sites proposed for development in the dolomite karst regions of southern Africa. In *Karst hydrology: Engineering and environmental applications: Proceedings of the second multidisciplinary conference on sinkholes and the engineering and environmental impacts of karst*, ed. B. F. Beck and W. L. Wilson, 331–36. Rotterdam: A. A. Balkema.

Rupert, F. R. 1991. *Lithology and palynology of cave floor sediments cores from Wakulla Spring, Wakulla County, Florida*. Open File Report no. 47. Tallahassee: Florida Geological Survey.

Ryan, C. R. 1984. High-volume grouting to control sinkhole subsidence. In *Sinkholes: Their geology, engineering and environmental impact: Proceedings of the First Multidisciplinary Conference on Sinkholes and the Engineering and Environmental Impacts of Karst, Orlando, Florida, 15–17 October 1984*, ed. B. F. Beck, 413–17. Rotterdam: A. A. Balkema.

Ryan, K. W. 1989. Two-dimensional numerical simulation of the relationship between sinkhole lakes and ground water along the Central Florida ridge. In *Engineering and environmental impacts of sinkholes and karst: Proceedings of the Third Multidisciplinary Conference on Sinkholes and the Engineering and Environmental Impacts of Karst*, ed. B. F. Beck, 33–45. Rotterdam: A. A. Balkema.

Salomone, W. G. 2007. The applicability of Florida mandatory endorsement for sinkhole collapse coverage: Part 1. Legal aspects. *Environmental Geology* 8:63–71.

Satchell, M. 1995. Sinkholes and stacks: Neighbors claim Florida's phosphate mines are a hazard. *U.S. News and World Report*, June 12, 53–57.

Schiffer, D. M. 1996. *Hydrology of the Wolf Branch Sinkhole Basin, Lake County, east-central Florida*. Open File Report no. 96-143. Tallahassee: U.S. Geological Survey.

Schmidt, W. 1997. Geomorphology and physiography of Florida. In *The geology of Florida*, ed. A. F. Randazzo and D. S. Jones, 1–12. Gainesville: University Press of Florida.

Schmidt, W., and T. M. Scott. 1984. Florida karst—Its relationship to geologic structure and stratigraphy. In *Sinkholes: Their geology, engineering and environmental impact: Proceedings of the First Multidisciplinary Conference on Sinkholes and the Engineering and Environmental Impacts of Karst, Orlando, Florida, 15–17 October 1984*, ed. B. F. Beck, 11–16. Rotterdam: A. A. Balkema.

Schoolman, M. 2001. *Reason and horror: Critical theory, democracy, and aesthetic individuality*. New York: Routledge Press.

Schueler, T. 1994. The importance of imperviousness. *Watershed Protection Techniques* 1(3): 100–11.

Scott, T. M. 1997. Miocene to Holocene history of Florida. In *The geology of Florida*, ed. A. F. Randazzo and D. S. Jones, 57–67. Gainesville: University Press of Florida.

Seale, L., L. Florea, H. Vacher, and R. Brinkmann. 2007. Using ALSM to map sinkholes in the urbanized, covered karst of Pinellas County, Florida—1. Methodological considerations. *Environmental Geology* 54(5):901–1046.

Shinn, E. A., C. D. Reich, S. D. Locker, and A. C. Hine. 1996. A giant sediment trap in the Florida Keys. *Journal of Coastal Research* 12(4): 853–59.

Sinclair, W. C., and J. W. Stewart. Sinkhole type, development, and distribution in Florida. Map prepared by the U.S. Geological Survey, Florida Department of Environmental Regulation, and Florida Department of Natural Resources. Florida Department of Environmental Protection. Website. http://www.dep.state.fl.us/geology/publications/sinkholetype3.pdf.

Sinclair, W. C., J. W. Stewart, R. K. Knutilla, A. E. Gilboy, and R. L. Miller. 1985. *Types, features, and occurrence of sinkholes in the karst of west-central Florida*. Water Resources Investigations Report no. 85-4126. Tallahassee: U.S. Geological Survey.

Sowers, G. F. 1984. Correction and protection in limestone terrain. In *Sinkholes: Their geology, engineering and environmental impact: Proceedings of the First Multidisciplinary Conference on Sinkholes and the Engineering and Environmental Impacts of Karst, Orlando, Florida, 15–17 October 1984*, ed. B. F. Beck, 373–78. Rotterdam: A. A. Balkema.

Smith, D. W., and A. F. Randazzo. 1987. Application of electrical resistivity methods to identify leakage zones in drained lakes. In *Karst hydrogeology: Engineering and environmental applications*, ed. B. F. Beck and W. L. Wilson, 227–34. Rotterdam: A. A. Balkema.

Snyder, S. W., M. W. Evans, A. C. Hine, and J. S. Compton. 1989. Seismic expression of solution collapse features from the Florida Platform. In *Engineering and environmental impacts of sinkholes and karst: Proceedings of the Third Multidisciplinary Conference on Sinkholes and the Engineering and Environmental Impacts of Karst*, ed. B. F. Beck, 281–98. Rotterdam: A. A. Balkema.

Spencer, S. M., and E. Lane. 1995. *Florida sinkhole index.* Open File Report no. 58. Tallahassee: Florida Geological Survey.

Sputo, T. 1993. Sinkhole damage to masonry structure. *Journal of Performance Constructed Facilities* 7(1):67–72.

Stangland, H. G., and S. S. Kuo. 1987. Use of ground penetrating radar techniques to aid in site selection for land application sites. In *Karst hydrogeology: Engineering and environmental applications*, ed. B. F. Beck and W. L. Wilson, 171–77. Rotterdam: A. A. Balkema.

Stewart, J. W. 1987. Potential recharge and ground-water contamination from selected sinkholes in west-central Florida. In *Karst hydrogeology: Engineering and environmental applications*, ed. B. F. Beck and W. L. Wilson, 247–52. Rotterdam: A. A. Balkema.

Stewart, J. W., and L. R. Mills. 1984. *Hydrogeology of the Sulphur Springs area, Tampa, Florida.* Water Resources Investigations Report no. 83-4085. Tallahassee: U.S. Geological Survey.

Stewart, M., D. Lei, R. Brinkmann, R. Aangeenbrug, and S. Dunlap. 1995. *Mapping of geologic and hydrologic features related to subsidence-induced foundation failures, Pinellas County, Florida.* Report prepared for Pinellas County Government.

Stewart, M., and J. Wood. 1984. Geophysical characteristics of fracture traces in the carbonate Floridan Aquifer. In *Sinkholes: Their geology, engineering and environmental impact: Proceedings of the First Multidisciplinary Conference on Sinkholes and the Engineering and Environmental Impacts of Karst, Orlando, Florida, 15–17 October 1984*, ed. B. F. Beck, 225–29. Rotterdam: A. A. Balkema.

Stone, P. A., P. J. Gleason, and G. L. Chmura. 2002. Bayhead tree islands on deep peats of the northeastern Everglades. In T*ree islands of the Everglades*, ed. F. H. Sklar and A. Van Der Valk, 71–115. Heidelberg: Springer.

Tanner, W. F., S. Demirpolat, and L. Alvarez. 1989. The "Gulf of Mexico" Late Holocene sea level curve. *Transactions of the Gulf Coast Association of Geological Societies* 39:553–562.

Thorp, M. J. W., and G. A. Brook. 1984. Application of double Fourier series analysis to ground subsidence susceptibility mapping in covered collapse terrain. In *Sinkholes: Their geology, engineering and environmental impact: Proceedings of the First Multidisciplinary Conference on Sinkholes and the Engineering and Environmental Impacts of Karst, Orlando, Florida, 15–17 October 1984*, ed. B. F. Beck, 197–200. Rotterdam: A. A. Balkema.

Tihansky, A. B. 1999. Sinkholes, west-central Florida. U.S. Geological Survey Circular. http://pubs.usgs.gov/circ/circ1182/pdf/15WCFlorida.pdf.

Tolmachev, V., and M. Leonenko. 2011. Experience in collapse risk assessment of building on covered karst landscapes in Russia. In *Karst Management*, ed. P. E. Van Beynen, 75-102. Heidelberg: Springer.

Troester, J. W., E. L. White, and W. B. White. 1984. A comparison of sinkhole depth frequency distributions in temperate and tropical karst regions. In *Sinkholes: Their geology, engineering and environmental impact: Proceedings of the First Multidisciplinary Conference on

Sinkholes and the Engineering and Environmental Impacts of Karst, Orlando, Florida, 15–17 October 1984, ed. B. F. Beck, 65–73. Rotterdam: A. A. Balkema.

Trommer, J. T. 1987. *Potential for pollution of the Upper Floridan Aquifer from five sinkholes and an internally drained basin in west-central Florida*. Water Resources Investigations Report no. 87-4013. Tallahassee: U.S. Geological Survey.

———. 1992. Effects of effluent spray irrigation and sludge disposal on ground water in a karst region, northwest Pinellas County Florida. Water Resources Investigations Report no. 97-4181. Tallahassee: U.S. Geological Survey.

Upchurch, S. B., and F. W. Lawrence. 1984. In *Sinkholes: Their geology, engineering and environmental impact: Proceedings of the First Multidisciplinary Conference on Sinkholes and the Engineering and Environmental Impacts of Karst, Orlando, Florida, 15–17 October 1984*, ed. B. F. Beck, 23–28. Rotterdam: A. A. Balkema.

Upchurch, S. B., and J. R. Littlefield. 1987. Evaluation of data for sinkhole-development risk models. In *Karst hydrogeology: Engineering and environmental applications*, ed. B. F. Beck and W. L. Wilson, 359–64. Rotterdam: A. A. Balkema.

———. 1988. Evaluation of data for sinkhole-development risk models. *Environmental Geology* 12(2):135–40.

Upchurch, S. B., and A. F. Randazzo. 1997. Environmental geology of Florida. In *The geology of Florida*, ed. A. F. Randazzo and D. S. Jones, 217–50. Gainesville: University Press of Florida.

U.S. Army Corps of Engineers (USACE). 1987. *Corps of Engineers wetland delineation manual*. Vicksburg, Miss.: USACE.

U.S. Army Corps of Engineers and South Florida Water Management District. 1999. Central and southern Florida project comprehensive review study: Final integrated feasibility report and programmatic environmental impact statement April 1999. Jacksonville, Fla.: USACE., Jacksonville District. http://www.evergladesplan.org/docs/comp_plan_apr99/summary.pdf.

U.S. Geological Survey. 2000. Borehole radar methods. http://water.usgs.gov/ogw/bgas/publications/FS-054-00/

U.S. Geological Survey Standing Scientific Group. 1996. South Florida ecosystem restoration: Scientific information needs. Science Subgroup Report to the Working Group of the South Florida Ecosystem Restoration Task Force. http://sofia.usgs.gov/publications/reports/sci_info_needs/.

Vacher, H., L. Seale, L. Florea, and R. Brinkmann. 2007. Using ALSM to map sinkholes in the urbanized covered karst of Pinellas County, Florida—2. Accuracy statistics. *Environmental Geology* 54(5):901–1046.

Van Beynen, P., N. Feliciano, L. North, and K. Townsend. 2007. Application of a karst disturbance index in Hillsborough County, Florida. *Environmental Management* 39(2):261–77.

Van Beynen, P., and K. Townsend. 2005. A disturbance index for karst environments. *Environmental Management* 36(1):101–16.

Van Der Valk, A., and F. Sklar. 2002. What we know and should know about tree islands. In *Tree islands of the Everglades*, ed. F. H. Sklar and A. Van Der Valk, 499–522. Heidelberg: Springer.

Veni, G. 1999. A geomporphological strategy for conducting environmental impact assessments in karst areas. *Geomorphology* 31(1): 151-80.

Vernon, R. Q. 1951. *Geology of Citrus and Levy Counties, Florida*. Bulletin no. 33. Tallahassee: Florida Geological Survey.

Villard, P., J. P. Gourc, and H. Giraud. 2000. Geosynthetic Reinforcement Solution to Prevent the Formation of Localized Sinkholes. *Canadian Geotechnical Journal* 37:987–99.

Votteler, T. H., and T. A. Muir. 2002. *Wetland management research wetland protection legislation.* U.S. Geological Survey Water Supply Paper no. 2425.

Wallace, R. E. 1993a. Dye trace and bacteriological testing of sinkholes: Sulphur Springs, Tampa, Florida. *Environmental Geology* 22:362–66.

———. 1993b. Dye trace and bacteriological testing of sinkholes tributary to Sulphur Springs, Tampa, Florida. In *Applied karst hydrology: Proceedings of the fourth multidisciplinary conference on sinkholes and the engineering and environmental impacts of karst*, ed. B. F. Beck, 89–96. New York: Taylor and Francis.

Waltham, T., F. Bell, and M. Culshaw. 2005. *Sinkholes and subsidence: Karst and cavernous rocks in engineering and construction.* Berlin: Springer.

West, F. E. 1973. *The Devil's Millhopper: A resource for developing field studies.* Gainesville, Fla.: P. K. Yonge Laboratory School.

Wetzel, P. R. 2002. Analysis of tree island vegetation communities: Hydrologic and fire impacts over a decade. In *Tree islands of the Everglades*, ed. F. H. Sklar and A. Van Der Valk, 357–89. Heidelberg: Springer.

White, W. A. 1970. *The geomorphology of the Florida Peninsula.* Geological Bulletin no. 51. Tallahassee: Florida Department of Natural Resources Bureau of Geology.

Whitman, D., T. L. Gubbels, and L. Powell. 1999. Spatial interrelationships between lake elevations, water tables, and sinkhole occurrence in central Florida: A GIS approach. *Photogrammetric Engineering and Remote Sensing* 65(10):1169–178.

Willard, D. A., C. W. Holmes, M. S. Korvela, D. Mason, J. B. Murray, W. H. Orem, and D. T. Towles. 2002. Paleoecological insights on fixed tree island development in the Florida Everglades: 1. Environmental controls. In *Tree islands of the Everglades*, ed. F. H. Sklar and A. Van Der Valk, 117–51. Heidelberg: Springer.

Wilson, K. V. 2004. Modification of karst depressions by urbanization in Pinellas County, Florida. Master's thesis, University of South Florida, Tampa.

Wilson, W. L., and B. F. Beck. 2005. Hydrogeologic factors affecting new sinkhole development in the Orlando area. *Ground Water* 30(6):918–30.

Wu, Y., K. Rutchey, W. Guan, L. Vilchek, and F. H. Sklar. 2002. Spatial simulations of tree islands for Everglades restoration. In *Tree islands of the Everglades*, ed. F. H. Sklar and A. Van Der Valk, 469–98. Heidelberg: Springer.

Zhou, W., and B. F. Beck. 2007. Management and mitigation of sinkholes on karst lands: An overview of practical applications. *Environmental Geology* 55:837–51.

Zisman, E. D. 2001. A standard method for sinkhole detection in the Tampa, Florida, area. *Environmental and Engineering Geoscience* 7:31–50.

Index

Page numbers in italics refer to illustrations.

Aerated caves, 191–92
Aerial photo and satellite analysis, 147, *148*, 149–50
Aetna Insurance, 162
Age of sinkholes, 221
Airborne laser swath mapping (ALSM), 141; accuracy, 152–53; expense, 152–53; filtration, 151–52; how it works, 150–52; new possibilities, 153–54; problems, 153
Alachua Tradition, 110–11
Almaleh, L. J., 89, 212
ALSM. *See* Airborne laser swath mapping
Alverson, D. C., 51
Anderson, L., 109
Apalachicola delta, 33; rarity of sinkholes in, 40–41
Appalachian Mountains, 23, 33
Aquifer, 13, 90, 100. *See also* Floridan Aquifer
Arcadia Formation, 118
Archival information, 155
Armentano, T. V., 103–4, *105*
Army Corps of Engineers, 93, 95, 185–86
Arrington, D. V., 77, 78
Arthur, J. D., 87
Attorneys, 9, 196
Avon Park Formation, 25, 44, 89

Bahtijarevic, A., 64
Banking, 188
Barfus, B. L., 49
Bartram, William, 112–13
Bear Sink, 76
Beck, Barry, 10–11, 55–56, 122–24, 147, 158
Bedrock, 72–73, 85–87; buoyancy, 45; covered, 49, 53, 138, 218; density, 74; Everglades, 38; geology defining, 220; horizontal, 19–20; limestone, 41, 51, 98, 102, 203–4; map, 211; pores, 27–28; structural characteristics, 64, 77, 142; subsurface, 30, 36, 56, 212; thickness, 97; voids, 29; young, 218
Bengtsson, Terrance, 44–46, 54
Benson, R. C., 89
Berg, R. R., 210

Bergado, D. T., 210
Best management practices (BMP), 183–84
Big Bend region, 87–88, *88*
Biscayne Aquifer, 90, 100
Bloomberg, D., 50, 51
Blue Pond, 63
Blue Sink, 59
BMP. *See* Best management practices
Bonaparte, R., 210
Bone Valley Formation, 23, 44, 66
Boring analysis, in geotechnical evaluation process, 199–200
Bowlegs, 113
Briar Sink, 76
Brinkmann, R., 56, 75
Brook, G. A., 63–64
Brooksville Ridge, 36–37, 98; sinkholes, 72–78
Broward, Napoleon Bonaparte, 92
Buffer zones, 188
Building codes, 168
Building collapse, *2*
Buried organic matter, 164
Bush, Jeffrey, 1, 2, 6

Calcite, 19, 74
Carr, W. J., 51, 103
Carson, Rachel, 94
Caverns, 39; Florida Caverns State Park, 36, 190; Floridan Aquifer, 59
Caves, 30, 75, 161, *191*; aerated, 191–92; collapse, 192; flooded, 191; grottos, 8, 14, 193; Mammoth Cave, 39; National Cave and Karst Research Institute, 10; policy, 190–93; vandalism, 192, *192*
Cenozoic era, 23–24, 33, 72, 73, 97, 98
Central Highlands, 34
Central Zone, 33
Chattahoochee Anticline, 40
Chenopodiaceae, 82–83
Circularity, 77
Citizens Insurance, 172–73
Citronelle Formation, 40, 77

Clays, 203–4; Hawthorne Formation, 140, 164; shrink-swell, 126, 135, 163–64; Tampa sinkhole swarm, 51–52
Clean Air Act, 94
Clean Water Act, 94, 182, 185
Climate variations, 17
Coal, 23
Coalescing sinkholes, 50
Coastal Lowland, 34
Coastal processes, *33*, 33–35
Cockpit karst landforms, 73
Cody Escarpment, 86
Coffee Sink, 76
Cold season pumping, 44–47
Collapse: building, *2*; caves, 192; cover collapse sinkholes, *29*, 29–30, 56, 60, 122–24; failure, 78; insurance sudden settlement or collapse phrase, 162–63; Lee Roy Selmon Expressway, 175–17, 175–76, *176*, 210; pipe, 214–15; pipeline collapse sinkholes, 215; sewer pipeline collapse sinkholes, 215; sinkholes, 1, 45, *48*, 178, 201, 208, 211; voids, 218; water pipeline collapse sinkholes, 215
Comprehensive Everglades Restoration Plan, 95
Conduit fill, 51
Cone penetration, 50
Cooling towers, 88–89
Corp of Engineers Wetland Delineation Manual, 186
Cover collapse sinkholes, *29*, 29–30, 56, 60, 122–24
Covered bedrock, 49, 53, 138, 218
Cover subsidence sinkholes, 29, *29*
Cowcatcher, 112–13
Cratons, 21
Crews Lake Sink A, 76
Crews Lake Sink B, 76
Critical Ecosystems Studies Initiative Science Subgroup, 95–97
Crooked Lake bathymetric survey, 65–66
Cross Bar Ranch, 57
Crystal River, 57, 212
Crystal Springs, 38, 88–89
Curiosity Sink, 59, 130
Currin, J. L., 49
C. W. Bill Young Regional Reservoir, 190
Cypress domes, 36, *38*, 63, 100, *187*
Cypresshead Formation, 25

Davis, William Morris, 63
Decision-making, 178
Department of Environmental Protection, 12
Depression contours, 156
Desalination, 46, 189
De Soto, Hernando, 111
Detection and mapping, 4; aerial photo and satellite analysis, 147, *148*, 149–50; ALSM, 141, 150–54; bedrock map, 211; field observation, 135–36; first statewide map, 157–58; GIS for, 141–42, 144–47; GPR, 136–40, *139*; gray literature and, 142–43; 145–46; historic records, 154–55; LIDAR mapping, 72, 151; map analysis, 141–47, *143*; overview, 134–35; private, 159; problems with, 134, 141, 145–56, 156–59; regional, 158; status of, 155; Tampa sinkhole swarm, 43; technology, 218–19; United Kingdom void map, 213–14, 221. *See also* Geologic map
Development rules, 174–80
Devil's Millhopper sinkhole, 3–4, 115–17, *116*
Disston, Hamilton, 91–92
Distribution of sinkholes, 31, 34–41, 157–58, 171
Dolina, 27
Doline, 27
Douglas, Marjory Stoneman, 90, 93
Drainage basins, 9
Drought, 133, 189
Dumps, 8, 14, 58. *See also* Landfills
Duricrusts, 98–99
Dye tracing, 130

Eastman, K. L., 171–72
Economic loss, 8. *See also* Property
EG&G UNIBOOM seismic system, 65
Endangered species protection, 9
Engineering karstology, 210
Engineering News Record, 214
Environmental Protection Agency (EPA), 185
Eocene epoch, 24, 72
Eogenetic karsts, 27–28
EPA. *See* Environmental Protection Agency
Evaluations and repairs, 220; arbitration over, 196; building on and in karst, 210–14; fighting evaluation, 196; geotechnical evaluation process, 197–202; geotechnical report interpretation, 202–7; insurance claims and, 194–97; nonsinkholes, 214–15; opting for repairs, 197; overview, 194; settling over, 196; stabilizing ground and saving structure, 207–9; total loss, 197
Everglades, 33, *38*, 73, 218; bedrock, 38; Comprehensive Everglades Restoration Plan, 95; drainage, 92, 104; geology, 97–100; history, 91–92; human modification in, 91, 104, 106; hurricanes, 92–93; hydrogeology, 99–100; pollution, 93–94; reclamation, 94–97; South Florida sinkholes, 90–100; tree islands, 100–106
Everglades: River of Grass, The (Douglas), 93

Faulting, 20
Fellowship soils, 125
FGS. *See* Florida Geological Survey
Field analysis, in geotechnical evaluation process, 199
Field evidence, 204
Field observation, 135–36
Flooded caves, 191
Flooding, 130–31
Flora and fauna, 12

Florea, L., 75
Florida: geology time scale, 21–24; peninsular, 25–26; satellite image, *35*, 35–36; unique landscape, 217. *See also* Geomorphology; Panhandle
Florida Atlas of Lakes, 63
Florida Caverns State Park, 36, 190
Florida Department of Financial Services, 12
Florida Geological Survey (FGS), 10, 40, 129, 167, 224; database, 147, 155, 158, 180, 198; founding and mission, 11–12; *Geologic and Geotechnical Assessment for the Evaluation of Sinkhole Claims*, 216; website, 11–12, 158, 219
Florida Keys, 89
Floridan Aquifer, 40, 44; caverns, 59; connections, 76; pollution, 57, 59; wells in, 52–53; Winter Park Sinkhole and, 108
Florida Natural History Museum, Gainesville, 23
Florida Sinkhole Index, 40
Florida Sinkhole Research Institute, 3, 5, 16, 53, 55, 122; closing, 11, 216; database, 147, 155, 158, 180; founding, 10; work done, 217
Florida State University, 11
Food Security Acts, 185
Formations: Arcadia, 118; Avon Park, 25, 44, 89; Bone Valley, 23, 44, 66; Citronelle, 40, 77; Cypresshead, 25; Ocala, 21, 44; Peace River, 25, 66, 118; Pliocene Citronelle, 25; Suwannee, 44; Tampa, 44; Torreya, 25. *See also* Hawthorne Formation
For the Future of Florida: Repair the Everglades, 94
Fossils, 23, 74
Frank, E. F., 122
French and Indian Wars, 112
Fretwell, J. D., 76

Garlanger, John E., 209, 211
Geographic Information System (GIS), 124–25, 142, 220; insurance and, 146–47; for map analysis, 144–47
Geologic and Geotechnical Assessment for the Evaluation of Sinkhole Claims, 216
Geologic map, 24; panhandle, 24–25; peninsular Florida, 25–26
Geologic setting: geomorphology of Florida, 32–34; overview, 31–32; sinkhole distribution, 34–41. *See also* Orlando sinkhole swarm; Tampa sinkhole swarm
Geologists, 167
Geology, 11; bedrock defined, 220; Everglades, 97–100; geologic cores, 20; Geologic Time Scale, 21, *22*; in geotechnical report interpretation, 203–4; time scale of Florida, 21–24
Geomorphology, 11, 15, 32–34, 163, 220
Geophysical investigations: in geotechnical evaluation process, 200–202; in geotechnical report interpretation, 205–6
Geotechnical engineer, 167
Geotechnical evaluation process: boring analysis, 199–200; conducting, 197; field analysis, 199; geophysical investigations, 200–202; previous evidence, 198–99
Geotechnical firms, 167–71
Geotechnical report interpretation: data interpretation, 207; executive summary, 202; field evidence at site, 204; geology in area, 203–4; geophysical investigations, 205–6; hand probes and, 204; mechanical probing, 206–7; sinkholes in area, 203
Gilboy, A. E., 51
GIS. *See* Geographic Information System
Global positioning system (GPS), 138, 145, 224
Goehring, R. L., 68
Golfball Sink, 76
GPR. *See* Ground-penetrating radar
GPS. *See* Global positioning system
Graham, Bob, 94
Grand Canyon, 31
Gray literature, 3, 14; map analysis, 142–43, 145–46
Green Swamp, 34, 54, 61
Grottos, 8, 14, 193
Ground instability, 55, 69–70, 164–65, 177, 200
Ground-penetrating radar (GPR), *139*, 201; cost-effectiveness, 219; equipment, 137–38; how it works, 136–37; improving and advancing, 224; overview, 136; problems, 140–41; processing, 138–40; readout, *205*; reports, 205–6
Groundwater pollution, 4, 86–87
Groundwater resources, 13, 46, 175
Groups: Hawthorne, 25, 80; Miocene Alum Bluff, 24; Sulphur Springs sinkhole, *129*, 129–31
Grouting, 208–9
Gypsum stacks, 47–49, *48*, 127–28, 223–24

Hall, L. E., 45
Hand probes, 204
Hansen, B. C. S., 66
Hawthornee Formation, 44, 50, 52, 64, 100; clay, 140, 164; North Florida sinkholes and, 85–86; permeability, 120; sediment, 98, 117–18; siliciclastic, 97–98
Hawthorne Group, 25, 80
Hazardous waste, 80
Hernando County Ordinance 94-8, 175
Hernasco Sink, 59, 76
Hildebrand, P. B., 71
Historic records, 154–55
Hollingshead, J. J., 85
Holocene epoch, 24, 99
Holszchuh, J. C., 46–47
Homosassa Springs, 89
Horizontal bedrock, 19–20
Human agency, 18
Hurricanes, 92–93, 128, 131
Hurston, Zora Neale, 92
Hydric soils, 149
Hydrogeology, Everglades, 99–100

Hydrology, 11, 31, 53, 60, 70, 178; altering, 92, 101, 114–15; interactive, 188; karst, 179; protecting, 179; stabilizing, 108; subsurface, 180; surface, 76; tree islands, 104; understanding, 97

IMC-Agrico, 128
Indurated or hard limestone, 25
Instability, 16, 174, 204; ground, 55, 69–70, 164–65, 177, 200; preventing, 208; structural, 195
Insurance, 4–5; adjusters, 168; Aetna Insurance, 162; buried organic matter and, 164; cancellations, 172–73; Center for Insurance Research, 11; Citizens Insurance, 172–73; costs, 170; in flux, 17–18; GIS and, 146–47; legal definition of sinkhole and, 163; paid advocates, 169; policy, 161–66; poor site preparation and, 165; raveling zones and, 165, *166*; shrink-swell clays and, 163–64; sinkhole activity and, 165; sinkhole claims, 7, 171–73, *172*, 181–82, 194–97, 222–23; State Insurance Commission, 170; state statute, 161–62; sudden settlement or collapse phrase, 162–63
Intercoastal Waterway, 182
Interlachen area, 77–78
Interpretation, 169–70
Interwellfield Project Area, 51

Jacksonville basin, 41, 98
Jammal, S. E., 108–9
Jensen, J. H., 83
Joint patterns, 50, 74, 86
Journal of Performance of Constructed Facilities, 215
Juniper Springs, 38

Karst, 3; areas, 9; building on and in, 210–14; carrying capacity, 179; cockpit landforms, 73; depressions, 27; eogenetic, 27–28; features, 11, 36–38, 73, 90, 211, 212; hydrology, 179; impact, 223; landforms, 15, 77; landscape, 28–29; National Cave and Karst Research Institute, 10; need for knowledge, 16; plains, 73; processes, 22, 24, 32, 177; research, 14; systems, 10; telogenetic, 27–28; terrain, 83; underground, 38; zoning, 5
Karstification, 29, 65, 73, 98
Kelly, Raymond, 160, 161
Kemmerly, P. R., 177–78, 180
Kentucky sinkholes, *15*, 174–75
Kindinger, J. L., 62, 84
Kissengen Springs, 46
Kissimmee River, 91, 95
Klein, K., 46
Klimchouk, A. B., 50
Kuhns, G. L., 70–71, 212
Kuo, S. S., 10-11, 71

Lacustrine wetlands, 186
Lake Jackson, 63
Lake Jessup, 63
Lake Kissimmee, 34, 61
Lake Magnolia, 63
Lake Okeechobee, 34, 91–93, 98, 127
Lakes: *Florida Atlas of Lakes*, 63; Orlando Lake District, 34–36; Orlando sinkhole swarm, 62–63; sediment, 65; Tampa Lake District, 34–36; Tampa sinkhole swarm, 52–53, *61*. See also *specific lakes*
Lake Wales Ridge, 36
Land, L. A., 89–90
Landfills, 8, 87
Land subsidence, 135–36
Land-use practice, 178–80
Lane, E., 38–39
Lawrence, F. W., 86
Lee Roy Selmon Expressway collapse, 175–76, *176*, 210
Lexington, Kentucky, sinkhole ordinance, 174–75
LIDAR mapping, 72, 151
Limestone: bedrock, 41, 51, 98, 102, 203–4; formation, 19–20; indurated or hard, 25; layers, 6; Ocala, 25; Oligocene Suwannee, 24; Quaternary, 25; Suwannee, 25
Lindquist, R. C., 77, 78
Littlefield, J. R., 53–54, 56
Loose zone, 70–71
Loss of life, 8
Lowe, D., 50

Magotes, 73
Mammoth Cave, 39
Map analysis, 141–47; disclosure, 144; electronic, 142; GIS for, 144–47; gray literature, 142–43, 145–46; limitations, 141; morphometric, 142, *143*; scale of map, 142–43
Marianna Lowlands, 34
Marine sand, 19–20, 69, 72–73, 203
Marine sediment, 28, *33*, 60, 218
Mechanical probing, 206–7
Menéndez de Avilés, Pedro, 111
Menéndez Marqués, Francisco, 111–12
Mesozoic era, 23
Metcalf, S. J., 45
Middle Miocene, 41
Miocene Alum Bluff Group, 24
Miocene epoch, 24, 97
Miocene sediment, 88
Mitigation projects, 187–88
Morphometric analysis, 142, *143*
Mount Lassen, 31
Muck, 70
Mulberry Phosphate Museum, 23

National Cave and Karst Research Institute, 10
National Speleological Society, 193
National Weather Service, 149
National Wetlands Inventory, 124–25, 126

Native Americans, 101, 109–11, 113–14; Seminole Wars, 91, 113–14
Natural disasters, 6, 8
Nature, fear of, 5–6, 8
Neogene period, 24
Newnan, Daniel, 113
New Wales Gypsum Stack Sinkhole, 127–28
Niagara Dolomite, 22
Nonsinkholes, 214–15
Northern Zone, 33
North Florida sinkholes: Big Bend region, 87–88, *88*; cooling towers, 88–89; density and depth, 79–80; environmental problems, 80; groundwater pollution, 86–87; Hawthorne Formation and, 85–86; joint patterns, 86; landfill problems, 87; poljes, 83–84, *84*; solution valleys, 80–81; Teagues Sinkhole, 85; urbanization induced, 84–85; Wakulla Springs, *81*, 81–83

Ocala Formation, 21, 44
Ocala Limestone, 25, 117
Ocala National Forest, *75*, 76
Ocala Ridge, 38, 98; sinkholes, 72–78
Oligocene epoch, 24
Oligocene Suwannee Limestone, 24, 118
Orange Lake, 63
Orange Springs, 76
ORE GEOPULSE seismic system, 65
Orlando Lake District, 34–36
Orlando sinkhole swarm: Crooked Lake bathymetric survey, 65–66; Forest City density, 64–65; lakes, 62–63; overview, 60–61; pollution and, 67–68, 71; radiocarbon tests, 66–67; stabilization projects, 68–70; subsurface testing, 69–70; trough directions, 64
Oros, R., 71
Ozark Plateau, 33

Padgett, D. A., 78
Paleocene epoch, 24, 97
Paleogene period, 24, 80
Paleogene sediment, 80
Paleokarst, 36, 65
Paleo-sinkholes, 76, 79, 86–87, 218
Paleozoic era, 21–22
Palustrine wetlands, 186–87
Panhandle: rarity of sinkholes in, 40; sediment, 24–25; sinkholes, 36
Patton, T. H., 46–47
Paull, C. K., 89–90
Payne, King, 113
Paynes Prairie sinkhole, 3–4, 8, *110*; access, 109; formation, 114; plugging and unplugging, 115; prairie history, 109–15
Peace River floodplain, 46–47
Peace River Formation, 25, 66, 118
Peninsular Florida, 25–26

Phosphate, 23, 47
Piers, 211–12
Pinellas County, *118*
Pinellas County sinkholes, 3–4, 55–56
Pinellas County sinkhole swarms, *121*; aerial photos, 119–20; cover collapse sinkholes, 122–24; drawdown of water, 124; foundation failures, 124–26; overview, 117–19; property damage, 119, 123; water-retention ponds, 120
Pipe collapses, 214–15
Pipeline collapse sinkholes, 215
Piping, 78–79
Pleistocene epoch, 24, 72, 82–83, 98, 99
Pleistocene sediment, 36, 72, 211
Pliocene Citronelle Formation, 25
Plugging, 114–15, 209
Policy, 4, 134, 219–20; cave protection, 190–93; Eastman study, 171–72; geotechnical firms, 167–71; Hernando County Ordinance 94-8, 175; insurance, 161–66; insurance cancellations, 172–73; lack of development rules, 174–80; land-use practice and, 178–80; Lexington, Kentucky, sinkhole ordinance, 174–75; overview, 160–61; questions, 17–18; real estate, 180–82; sinkhole industry, 166–71; stakeholders, 222; subsurface testing and, 175–78; water, 182–84; well fields, 189–90; wetlands, 185–89
Poljes, 63, 83–84, *84*
Pollution, 78, 130; Everglades, 93–94; Floridan Aquifer, 57, 59; groundwater, 4, 86–87; Orlando sinkhole swarm and, 67–68, 71; surface-water, 4; Tampa sinkhole swarm and, 57–59
Ponds: Blue, 63; Smokehouse, 76; water retention, 120, 122, 130, 141, 150, 183, *184*
Poor site preparation, 165
Porosity: sand, 204; sediment, 137
Precambrian eon, 21
Previous evidence, in geotechnical evaluation process, 198–99
Price, D. J., 80, 85–86
Private detection and mapping, 159
Property, 6, 10; house damage, 7, 18, 119, 123, 134, 160–63, 180–82; structural damage, 4, 5, 170, 195, 215. *See also* Real estate
Public perception, 5–8

Quarrying, 74–75
Quartz sand, 98, 147
Quaternary Limestone, 25
Quaternary period, 24, 25, 69
Quaternary sand, 79, 117–18, 140
Quaternary sediment, 25, 80
Queen Anne's War, 113

Radiocarbon tests, 66–67
Rainbow Springs, 38
Randazzo, A. F., 84–85

Raveling: sand, 162, *166*, 169; sediment, 29, 122, 165, 218; zones, 140, 165, *166*, 183, 199–200, 206
Real estate, 180–82; wetlands as, 186
Reeder, P., 75
Regional detection and mapping, 159
Reported sinkholes, 157
Research, 5, 10–11; enhancing, 224; karst, 14; proposed SRR, 216. *See also* Florida Geological Survey; Florida Sinkhole Research Institute
Ridges, *37*; Brooksville, 36–38, 72–78; formation and makeup, 72; Lake Wales, 36; Ocala, 38, 72–78, 98
Risk levels, 213
Risk models, 54–55, 222
Rivers: Crystal River, 57, 212; Kissimmee, 91, 95; Peace River floodplain, 46–47; Peace River Formation, 25, 66, 118
Roads, 210
Rock Sink, 76
Rocky Sink, 76
Round Sink, 76
Rubber sheet, 149
Rupert, F. R., 81–83
Ryan, K. W., 67–68

Sand, 25, 28, 41, 44, 60, 204; filtering, 6, 70, 203; marine, 19–20, 69, 72–73, 203; in pipes, 51, 87; porosity, 204; quartz, 98, 147; Quaternary, 79, 117–18, 140; raveling, 162, *166*, 169; settling, 160; unsaturated, 138
Save Our Everglades, 94
Sayed, S., 55–56, 68, 122–24
Schiffer, D. M., 67
Schmidt, W., 33–34, 40, 41, 62, 89
Schoolman, M., 5
Science questions, 16–17
Scott, T. M., 40, 41, 89
Sediment, 32, *166*, *191*; Citronelle Formation, 40, 77; cohesive, 158; complex, 4; cover, 49–50, 53–55, 76, 218; duricrusts, 99; filtering, 77–78, 120, 200; Hawthorne Formation, 98, 117–18; lake bottom, 65; marine, 28, *33*, 60, 218; Miocene, 88; noncohesive, 51; Paleogene, 80; Panhandle, 24–25; Pleistocene, 36, 72, 211; porosity, 137; Quaternary, 25, 80; raveling, 29, 122, 165, 218; shrinking and, 164; size, 204; slope wash and, 32; subsurface, 202, 207; surface, 23, 203; thickness, 26, 40–41, 43, 86, 97; unconsolidated, 70. *See also* Clays; Sand; Silt
Seffner Sinkhole, 2
Seminole Wars, 91, 113–14
Sewer pipeline collapse sinkholes, 215
Sewers, 212, 214–15
Shinn, E. A., 90
Shrink-swell clays, 126, 135, 163–64
Silent Spring (Carson), 94
Silt, 50–51, 66, 69, 89, 204
Sinclair, W. C., 60, 78, 157–58
Sinkhole formation, 16, 218; basic concepts, 19–21;

geologic map, 24–26, *26*; mechanism, 222; overview, 19; reasons for, 17; time scale of Florida geology, 21–24; types and definitions, 26–30; unknown factors, 221
Sinkhole research resource (SRR), 216
Sinkholes: activity, 165; age of, 221; areas where rare, 40–41; coalescing, 50; cold season pumping, 44–47; collapse, 1, 45, *48*, 178, 201, 208, 211; cover collapse, *29*, 29–30, 56, 60, 122–24; cover subsidence, 29, *29*; culture and, 32; damage, 6; defining, 156, 223; distribution, 31, 34–41, 157–58, 171, 218; expertise, 10–16; fear of, 6; gypsum stacks and, 47–49, *48*; human agency and, 18; industry, 166–71; insurance claims, 7, 171–73, *172*, 181–82, 194–97, 222–23; international terms, 27; landfills, 8, 87; legal definition, 163; paleo, 76, 79, 86–87, 218; Panhandle, 36; as parks, 8–9; pipeline collapse, 215; polje sinks, 63, 83–84, *84*; in popular culture, 3; regularity of, 2; reported, 157; risk models, 54–55; satellite image, *35*; sewer pipeline collapse, 215; shallow, 7; size and shape, 31–32; solution, 28; South Africa, 212; spray-effluent irrigation and, 47; standards summit, 167; trends, 17; types and definitions, 26–30; unreported, 155; water pipeline collapse, 215. *See also specific sinkholes; specific topics*
"Sinkhole Type, Development, and Distribution in Florida," 157–58
Slavery, 111, 113
Slope wash, 32
Smith, D. W., 84–85
Smokehouse Pond, 76
Snyder, S. M., 65, 80
Soil consistency, 204
Soil tests, 168
Solution sinkholes, 28
Solution valleys, 80–81
South Africa sinkholes, 212
Southern Zone, 33
South Florida Basin, 41
South Florida sinkholes: bridge building, 89–90; Everglades, 90–100; Everglades tree islands, 100–101; Florida Keys, 89; low risk zone, 89; tree island formation, 101–6
Sowers, George F., 210–11
Speleogenesis, 30
Speleology, 11, 74
Speleothems, 191–92
Spray-effluent irrigation, 47
Spring Hill Sinkhole, 3–4, 131–33, *132*
Springs, 131–33, 161; Crystal, 38, 88–89; Homosassa, 89; Juniper, 38; Kissengen, 46; Orange, 76; Rainbow, 38; Sulphur Springs Sinkhole, 3–4; Sulphur Springs sinkhole group, *129*; Wakulla, *81*, 81–83
SRR. *See* Sinkhole research resource
Stabilizing ground and saving structure, 207–9
Standardization, 167, 168

Stangland, H. G., 71
State Insurance Commission, 170
Stewart, J. W., 52–53, 124–25, 157–59
Stewart, M., 57
Stone, P. A., 103
Storm-water management, 183–84
Stratigraphic Code, 20–21
Stratigraphy, 20–21
Stratomax Sink, 76
Structural adequacy, 168
Structural engineer, 167
Structural instability, 195
Subsurface bedrock, 30, 36, 56, 212
Subsurface sediment, 202, 207
Subsurface testing, 69–70, 175–78
Subsurface voids, 208, 213–14
Sudden settlement or collapse, 162–63
Sulphur Springs Sinkhole, 3–4
Sulphur Springs sinkhole group, *129*, 129–31
Superposition, 20
Surface sediment, 23, 203
Surface-water pollution, 4
Surfer, 152
Suwannee Formation, 44
Suwannee Limestone, 25
Swallet, 27
Swamps, 23, 186; Green Swamp, 34, 54, 61; Swampbuster Act, 91, 185; tidal, 118

Tallahassee Hills, 34
Tampa Bay Water, 189–90
Tampa Formation, 44
Tampa Lake District, 34–36
Tampa sinkhole swarm: boundaries, 41–42, *42*; clays, 51–52; cold season pumping, 44–47; conduit fill, 51; detection and mapping, 43; dimensions and densities, 49; gypsum stacks, 47–49, *48*; identifying ancient and modern sinkholes, 53–55; joint patterns, 50; lakes, 52–53, *61*; Peace River floodplain, 46–47; Pinellas County, 42–43, 55–56; pollution and, 57–59; sediment cover, 49–50, 218; spray-effluent irrigation, 47; urbanization and, 56–57; as Zone 2 and Zone 5, 60
Teagues Sinkhole, 85
Telogenetic karsts, 27–28
Ten Thousand Islands, 38, *39*
Tertiary period, 24, 98
Their Eyes Were Watching God (Hurston), 92
Thomas, S. J., 215
Thorp, M. J. W., 63–64
Tihansky, Ann, 43–47, 49–50, 158
Topographic anomalies, 202
Torreya Formation, 25
Tree islands: classification, 103–4, *105*; Everglades, 100–106; floating, 101–2; formation in South Florida, 101–6; hydrology, 104; stationary, 102; vegetative communities, 104

Troester, J. W., 79
Troitzky, G. M., 210
Trommer, J., 57, 59
Trough directions, 64

United Kingdom void map, 213–14, 221
United States Deep Diving Team, 83
United States Geological Survey (USGS), 43, 53, 140, 145, 159, 219
University of Florida, 216
University of South Florida, 11, 51, 216
University of South Florida Sinkhole, 1
Unreported sinkholes, 155
Upchurch, S. B., 50, 53–54, 86
Urbanization: activities, 155; rapid, 150; Tampa sinkhole swarm and, 56–57; urban decay, 131
USGS. *See* United States Geological Survey
Uvalas, 27, 50, 73

Veni, George, 178–80
Voids, 210; bedrock, 29; collapse, 218; space, 78; subsurface, 208, 213–14; United Kingdom map, 213–14, 221

Wakulla Springs, *81*, 81–83
Wallace, R. E., 130
Water, 215; Clean Water Act, 94, 182, 185; drawdown in Pinellas County sinkhole swarms, 124; groundwater pollution, 4, 86–87; groundwater resources, 13, 46, 175; Intercoastal Waterway, 182; management districts, 13, *13*, 178, 188; policy, 182–84; recharge areas, 9; retention ponds, 120, 122, 130, 141, 150, 183, *184*; storm-water management, 183–84; surface-water pollution, 4; withdrawal, 221. *See also* Lakes; Ponds; Rivers; Springs; Swamps; Wetlands
Water pipeline collapse sinkholes, 215
Wells: fields policy, 189–90; in Floridan Aquifer, 52–53
Wetlands, 8, 32; banking, 188; biomass, 185; buffer zones, 188; delineation, 186–87; lacustrine, 186; laws, 35; mitigation projects, 187–88; National Wetlands Inventory, 125–26; palustrine, 186–87; policy, 185–89; as real estate, 186. *See also* Swamps
White, W. A., 33–34
Whitman, Dean, 159
Wilson, K. V., 43, 119–20, 188
Winter Park Sinkhole, 3–4, *7*, 68, 220; breaches, 108; catalyst for study, 217; Floridan Aquifer and, 108; formation, 5–6, 10, 108; location and site, 107; size, 108; stabilizing of, 108
Wolf Branch Sinkhole, 67
Wood, J., 57

Zimmer v. Aetna Insurance Company, 162
Zisman, E. D., 168–69

ROBERT BRINKMANN is director of sustainability studies and professor in the Department of Geology, Environment, and Sustainability at Hofstra University. He is also the director for sustainability research at the National Center for Suburban Studies at Hofstra University. Prior to taking his current position, he was professor at the University of South Florida. Brinkmann also chairs the board of the National Cave and Karst Research Institute.